"十四五"普通高等教育本科部委级规划教材
纺织科学与工程一流学科建设教材
纺织工程一流本科专业建设教材

纺织英语（第4版）

Textile English（4th edition）

刘 雍 主编

中国纺织出版社有限公司

内 容 提 要

本书包括纺织原料、纱线生产、织物生产、织物染整和织物检验等内容,同时还涉及非织造布、高性能纤维、纺织复合材料、服装等内容。本书每篇课文后均附有生词表、课文注释以及为理解课文内容而设置的问题。书后附有总单词表,在本书配套的数字化资源中提供了专业微课堂、思政微课堂和语言微课堂。通过学习本书,除了可以掌握纺织专业基本知识,还可以掌握纺织领域常用的英语词汇以及基本的纺织英语资料翻译技巧。

本书可作为普通高等院校纺织工程专业学生的教材,也可作为相关企事业单位和科研单位学习纺织英语的教材,同时也可提供给具有一定英语基础并且对纺织生产感兴趣的人员学习使用。

图书在版编目(CIP)数据

纺织英语/刘雍主编.--4版.--北京:中国纺织出版社有限公司,2021.8(2025.1重印)

"十四五"普通高等教育本科部委级规划教材 纺织科学与工程一流学科建设教材 纺织工程一流本科专业建设教材

ISBN 978-7-5180-8697-9

Ⅰ.①纺… Ⅱ.①刘… Ⅲ.①纺织—英语—高等学校—教材 Ⅳ.①TS1

中国版本图书馆CIP数据核字(2021)第138590号

责任编辑:沈 靖 孔会云 责任校对:王蕙莹
责任印制:何 建

中国纺织出版社有限公司出版发行
地址:北京市朝阳区百子湾东里A407号楼 邮政编码:100124
销售电话:010—67004422 传真:010—87155801
http://www.c-textilep.com
中国纺织出版社天猫旗舰店
官方微博 http://weibo.com/2119887771
三河市宏盛印务有限公司印刷 各地新华书店经销
2001年2月第1版 2005年8月第2版
2008年4月第3版 2021年8月第4版
2025年1月第4次印刷
开本:787×1092 1/16 印张:17.5
字数:363千字 定价:56.00元

凡购本书,如有缺页、倒页、脱页,由本社图书营销中心调换

前言(第4版)

一、基本情况

《纺织英语》(第4版)由天津工业大学刘雍教授组织相关授课教师团队,在黄故教授主编的《纺织英语》(第3版)的基础上修订而来。本教材主要用于高等学校纺织工程及相关专业的专业英语教学。

二、教材定位

《纺织英语》(第4版)修订过程中,严格遵照《教育部关于一流本科课程建设的实施意见》(教高〔2019〕8号)的要求,落实立德树人根本任务,紧紧围绕高校办学定位和人才培养目标定位,深入挖掘蕴含的思想政治教育元素,以建设适应新时代要求的一流本科课程为宗旨,确定了本书的定位、编写思路与特色。教材主要面向纺织类工科高等院校,结合课程思政元素,突出实用性,能满足线上与线下教学需求,在课程思政、课程设计、知识结构、能力素质等方面进行了重新设计与编排,对传统教学模式进行了改进,体现"新工科"和一流人才培养的需要。

三、教材内容与特色

修订时,教材内容坚持符合学生成长规律,依据纺织学科前沿动态与产业发展需求动态更新知识体系,体现思想性、科学性与时代性。主要内容与特色包括:

(1)落实课程思政。

为落实课程思政要求,在大多数课程内容中增加"思政微课堂",包括我国近代史上做出重要贡献的著名纺织专家和教育家张汉文、严灏景等人的事迹,体现他们的科学报国、家国情怀和担当精神;纺织史中突出了我国纺织对世界和人类文明所做的贡献,特别是古代的丝绸之路与现代的"一带一路";在棉纤维课文中增加了我国新疆棉的特点以及揭露国外有关新疆棉的谎言;在纺织检测中增加了对我国纺织检测标准的宣传;在防护纺织品中特别增加了我国对新冠肺炎疫情有效控制、我国防护纺织品对世界抗疫的贡献等,充分激发学生的民族自豪感和爱国热情,突出文化自信、制度自信和专业自信。"思政微课堂"专门制作了ppt,或采用相关视频,以二维码的形式在每课中呈现。

(2)优化课程设计。

课文篇目在保持涵盖纺织领域大多数内容的基础上,由第3版的70课减少到57课。为了照顾不同学校和不同层次学生的教学进度,建议部分课文内容为精读内容,部分为泛读或自学内容(带*的课文),各校授课时可根据实际情况和自身专业特色与优势进行必要取舍。

(3)优化知识结构。

对知识结构进行了优化。如第3版教材中关于棉纤维一共三课,本次修订整合为一课,对其余纤维及工艺章节和内容也进行了整合与精简,特别是对原教材中关于纺纱工艺、织造工艺、毛纺、染整、非织造材料等部分课文进行了合并。根据纺织史、纺织工业现状及纺织科技进展,增加了部分内容。如第一课增加了纺织史、纺织与现代生活和纤维分类等内容;考虑到纳米纤维发展迅速,专门增加了纳米纤维内容;针对病毒防护,增加了防护纺织品相关内容;也增加了纺织品检测标准、可穿戴电子智能纺织品和未来纺织等相关内容;另外,为更好地理解课文内容,在课文中增加了图片和示意图。

为体现英语应用特点,课文的选编多精选英美的原版教材、专著、期刊论文或政府机构网站等,力求语言流畅、语法规范,内容力求反映纺织工艺基本原理和相关科技最新进展。书中涵盖了纺织史、纤维、纱线、纺织工艺、面料、染整技术、非织造材料、时装、纺织复合材料、功能纺织品、智能纺织品等内容,并对未来纺织技术做了介绍,以专业为主,兼顾科普性、实用性和前沿性。书中包括了纺织专业常用的基本词汇。

(4)提供线上教学资源。

在常态化疫情防控背景下,在线教学逐渐成为与线下教学并行的教学方式。为配合在线教学需要,在每篇课文中,精选了重要段落,提供了在线学习的“语言微课堂”;对专业部分,精心制作和优选了视频资源作为“专业微课堂”,供在线教学和学生自学选用。学生通过扫描课文中二维码即可观看和学习。

语言微课堂演示

专业微课堂演示

四、致谢

本书修订过程中,刘雍负责本书第1、13、40、42~48、54~57课的修订,李辉芹负责第2~3、6~7、29~31课的修订,李翠玉负责第4~5、22~24课的修订,赵健负责第8~10、32~33、49课的修订,杨光负责第11~12、37~38课的修订,张美玲负责第14~21课的修订,陈磊负责第25~28、51~53课的修订,刘丽妍负责第34~36课的修订,厉宗洁负责第39、41、50课的修订,全书由刘雍统稿。

本书音频资料由天津工业大学人文学院赵晟(第1~25、28课)和杨惠君(第26~27、29~57课)录制。部分视频资料来源于百度好看视频、腾讯视频、优酷视频、哔哩哔哩网站、微博网等,在此我们对所有视频制作、发行方表示诚挚的感谢,所有视频仅用于教学,方便学生学习使用,如有疑问请与我们联系。本书部分资料也得到了美国阿肯色大学助理教授寿万博士的帮助,在此一并致谢。

本书可以作为普通高等院校纺织工程专业学生的教材,也可以作为纺织一流专业建设和新工科专业建设的教材,同时也可作为相关企事业单位和科研单位学习纺织英语的教材,也可供对纺织领域感兴趣的读者参考使用。

由于编者水平有限,书中难免有疏漏或错误之处,欢迎读者对本书提出宝贵意见和建议。

刘雍

2021年2月

Preface(4th edition)

Textile English(*4th edition*) is revised by Prof. Yong Liu and his colleagues from School of Textile Science and Engineering of Tiangong University, based on *Textile English* (*3rd edition*) edited by Prof. Gu Huang. This book is mainly used to the English reading course for textiles, clothing and related majors.

During revising this book, we followed the requirements of the "Opinions of the Ministry of Education on the Construction of First-class Undergraduate Courses" (Jiaogao [2019] No. 8) to practice the fundamental task of establishing moral integrity in cultivation, and closely focus on the administrative position of universities and the objective of talents training. After deeply excavating the elements of ideological and political education, and taking the construction of first-class undergraduate courses to meet the requirements of the new era as the purpose, we identified course objectives, the writing ideas and characteristics of this book. Combining with ideological and political education, the book is intended for the students in textile colleges and universities, satisfying demands for online collective teaching and offline classroom teaching. The ideological and political education, course design, knowledge structure and teaching mode were rebuilt and changed to meet the needs of cultivating high-caliber personnel for Emerging Engineering Education in China.

In order to reflect the ideological, scientific and contemporary character, the content of this book was designed to follow the law of students' growth, and update the knowledge system dynamically according to the frontier dynamics of textile discipline and the needs of social development. The main revisions of this book include but are not limited to:

(1) Implementing ideological and political education.

Adding the ideological and politicalmicro class (IPMC) in most chapters in this book. The stories and the spirits of utter devotion and serving the country through science of several famous textile experts and textile educators, e.g., Prof. Haojing Yan and Prof. Hanwen Zhang, in modern China. Remembering them means letting their spirits inspire us in our daily lives. The contribution of China's textiles to the world and human civilization, especially the ancient Silk Road and the modern "Belt and Road" are highlighted in the textile history content. The characteristics of Xinjiang cotton in China and revealing few western countries' lies on Xinjiang cotton are provided in the lesson of cotton. The textile and clothing inspection standards in China are described. In particularly, the effective control of Novel Coronavirus epidemic (COVID-19) in China and the contribution of China's protective textiles to the world's anti-epidemic are described in the protective textiles, highlighting our confidence in the socialist path, theories, system and culture of socialism with Chinese characteristics. We prepared coursewares or videos for IPMC showing QR codes in the text.

(2) Optimizing the curriculum design.

The number of lessons was reduced from 70 in the third edition book to 57 in the fourth edition including almost all the textile fields. In order to take care of the teaching schedules in different universities and students at different levels, it is suggested that part of the contents should be intensive reading, and part should be extensive reading or self-study content (lesson with *). The teachers can make their choices according to the actual situation and its own professional characteristics and advantages.

(3) Optimizing the knowledge structure.

Integration of some lessons from the original texts of the third edition book. For example, there is only one lesson about cotton in the revised book integrating from three lessons in the third edition. Other chapters and contents of fibers and technologies are also integrated and simplified according to teaching needs. Most lessons in the original book on spinning technology, weaving technology, wool spinning, dyeing and finishing, and nonwoven materials are integrated in this new version.

Adding several lessons according to the textile history, the current situation of the textile industry and the development of textile science and technology. For example, the first lesson is changed as the history of textile, textile and modern life and fiber classification. Considering the rapid development of nanofibers, the preparation and the application of nanofibers is added in the fourth edition. The knowledge of protective textiles is also added for virus protection. Textile testing standards, wearable electronic smart textiles and future textiles are also added in the revised book. In addition, in order to better understand the content of the text, some new pictures and schematic diagrams are listed in the text.

The text of this book is carefully selected mainly from textbooks, monographs, journals or websites of government agencies in China, the United Kingdom and the United States. The text reads easily and smoothly with standard expressions. The book covers the history of textile, fiber, yarn, textile technology, fabric, dyeing and finishing technology, nonwoven materials, fashion, textile composite materials, functional textiles, intelligent textiles, etc. It also introduces the future textile technology, focusing on science popularization, taking into account professionalism, practicability and frontier. The book includes the basic vocabularies commonly used in textile and clothing major.

(4) Providing the resources for online teaching.

Under the background of regular epidemic prevention and control, online teaching offline classroom teaching has gradually equal importance. We provided audio as language micro class for several paragraphs in every course in this book. Additionally, we also provided several professional coursewares or videos as textile micro class in every course in this book. Anyone can learn them through scanning QR codes in the text.

Thanks for all the contributors of the book. Prof. Yong Liu is responsible for the revision of Lessons 1, 13, 40, 42 – 48 and 54 – 57, Dr. Huiqin Li is responsible for Lessons 2 – 3, 6 – 7 and 29 – 31, Dr. Cuiyu Li is responsible for Lessons 4 – 5 and 22 – 24, Dr. Jian Zhao is responsible for Lessons 8 – 10, 32 – 33 and 49, Dr. Guang Yang is responsible for Lessons 11 – 12 and 37 – 38, Dr. Meiling Zhang is responsible for Lessons 14 – 21, Dr. Lei Chen is responsible for Lessons 25 – 28 and 51 – 53, Dr. Liyan Liu is responsible for Lessons 34 – 36, Dr. Zongjie Li is responsible for Lessons 39, 41 and 50.

Thanks for Sheng Zhao and Huijun Yang from School of Humanities in Tiangong University. They read and recorded all the audio in this book, e.g., Zhao for Lessons 1–25, 28 and Yang for Lessons 26–27, 29–57. Thanks for Internet for most videos. They include, but are not limited to, the following websites. http://haokan.baidu.com. https://v.qq.com/. www.youku.com. www.bilibili.com. www.weibo.com. We want to express our thanks to these publishers, distributors and producers for their work. All the videos will be only used in teaching and students' study. If the publishers and anyone have any question, please do not hesitate to contact with us. Thanks for Dr. Wan Shou from University of Arkansas Fayetteville, USA for his help during revising this book.

This book can be used as a textbook for students majoring in textile engineering in colleges and universities, as textbook for the construction of first–class textile major and new engineering major, also can be used as a textbook for related majors, public institutions and scientific research institutions to learn textile English, and as a reference for readers interested in the textile field.

Due to our limited ability and knowledge, however, it is inevitable that there are omissions or errors in the book. It is very much appreciated if readers could provide valuable opinions and suggestions on this book.

Yong Liu

Feb, 2021

前言(第1版)

本书是为高校纺织专业学生编写的专业英语教材。书中的课文均精选于英美的原版教材,语言流畅,语法规范。课文的内容涉及到纺织专业的各个领域,包括新型纺织技术以及现代纺织技术的介绍。书中涵盖了纺织专业常用的基本词汇。

本书尽管是为高校学生编写的,但也可以作为纺织企业技术人员英语的培训教材。本书也可供对纺织有兴趣的读者参考使用。

在编写本书的过程中,赵书林、宋广礼、李亚滨、吴关臣、刘建中、杨锁廷、郭秉臣等同志给予了极大的支持和帮助,在此一并表示诚挚的谢意。

热忱欢迎读者对本书提出宝贵意见。

黄故

2000年10月17日

Preface(1st edition)

The book is written for college students specialized in textiles. All the texts are carefully chosen from books published in the United States or United Kingdom. Each article reads easy and smooth with standardized expressions.

The context is concerned with various fields in textiles including the introduction of new textile technologies. Words selected in the book are sufficient for daily use in textile communication.

Although the book is prepared for college students, it is also suitable for English training programs for technicians and engineers in mills. The book is also of reference to readers who are interested in textiles.

The author would like to thank the invaluable support provided by Zhao Shulin, Song Guangli, Li Yabin, Wu Guanchen, Liu Jianzhong, Yang Suoting and Guo Bingchen in writting the book.

Comments and suggestions about the books are both welcomed.

Huang Gu
Oct. 17,2000

前言(第2版)

本书是《纺织英语》的第2版,是在第1版基础上改编而成的。主要适用于高等院校纺织及相关专业的专业英语教学。书中的课文均精选于英、美的原版教材或杂志,语言流畅,语法规范,内容力求反映纺织技术的最新发展。书中涵盖了纤维、纱线、织物、纺织品染整、非织造布、时装、复合材料等内容,并对新型纺织技术做了介绍。书中包括了纺织专业常用的基本词汇。

与第1版《纺织英语》相比,本书增加了较多的内容,因此更适于学生的英语水平。

本书尽管是为高校的学生编写的,但也可以作为相关企业学习英语的教材。对纺织领域有兴趣的读者也可以参考使用。

欢迎读者对本书提出宝贵意见和建议。

黄故

2004年10月17日

Preface(2ed edition)

The book is the second edition of *Textile Readers* which is based on it. The main purpose of the book is for the college students majoring in textiles and related majors. The context is carefully selected from books and journals published in the United States and the United Kingdom. It is easy and smooth for students to read the texts with standard expressions. The book includes fibers, yarns, fabrics, textile dyeing and finishing, nonwovens, fashion, composites; and new textile technology is also included to reflect the newest development in the textile field. Words selected in the book are sufficient for daily use on textile communication.

Compared with the late *Textile Readers*, the book is more substantial, it is sought that it may more suitable to the students.

Although the book is prepared for the college students, anyway, it can also be used as training materials for enterprises concerned. For those who are interested in the textiles may use the book as reference.

Comments and suggestions are welcomed on the book.

Huang Gu
Oct 17, 2004

前言(第3版)

本书是在《纺织英语》(第2版)的基础上改编而成的。主要用于高等院校纺织及相关专业的专业英语教学。书中的课文多精选于英美的原版教材或杂志,语言流畅,语法规范,内容力求反映纺织技术的最新发展。书中涵盖了纤维、纱线、织物、纺织品染整、非织造布、时装、复合材料等内容,并对新型纺织技术做了介绍。根据有关院校使用该教材的反映,对内容做了部分删改和补充。书中包括了纺织专业常用的基本词汇。

本书尽管是为高校纺织专业的学生编写的,但也可以作为相关企业学习英语的教材。对纺织领域有兴趣的读者也可以参考使用。

欢迎读者对本书提出宝贵意见和建议。

黄故

2008年2月

Preface (3rd edition)

The book is written based on the *Textile Readers* (*2nd edition*). It is used for the college students majoring in textiles and related majors. The context is carefully selected mainly from books and journals published in the United States and the United Kingdom. Each text reads easily and smoothly with standard expressions. The book covers fibers, yarns, fabrics, textile dyeing and finishing, nonwovens, fashion, composites, and new textile technology is also included, which reflects the newest development in the textile field. It is revised based on the suggestions raised from different readers. Words selected in the book are sufficient for daily use on textile communication.

Although the book is prepared for the college students, it can also be used as training materials for enterprises. Those who are interested in the textiles may use the book as reference.

All suggestions for improvement are cordially welcome.

Huang Gu

Feb, 2008

Contents
目　录

Lesson One

Textiles: History, Development and Category

Textiles and Modern Life

The term "Textile" is a Latin word originated from the word "texere" which means "to weave". Textile refers to a flexible material comprising of a network of natural or articial fibers. Textiles are formed by weaving, knitting, braiding, nonwovens and other methods of pressing fibers together.

The evolutionary pathway of textiles goes back to the Stone Age with men and women wore clothing made of animal skins and plant leaves to protect them from the cold, sun, sand, and dust, this can be considered as the birth of textiles (Fig.1.1). From the earliest hand-held spindle and distaff and basic handloom to the highly automated spinning machines and power looms of today, the principles of turning vegetable fiber into cloth have remained constant: Plants are cultivated and the fiber harvested. The fibers are cleaned and aligned, then spun into yarn or thread. Finally, the yarns are interwoven to produce cloth. Today we also spin complex synthetic fibers, but they are still woven together using the same process as cotton and flax were millennia ago.

Fig.1.1 Animal skins for prehistoric people

Nowadays, textiles are an integral part of our world. Textiles touch our daily lives, from casualwear to household textiles to more technically advanced materials used in medical applications to industrial products. When we wake up in the morning, we raise our head from pillows covered with fabric and often filled with fibers and climb out from under sheets and blankets. We step into slippers and slip into robes. We wash our bodies with washcloths and dry them with towels. We brush our teeth with toothbrushes; the bristles are synthetic textile fibers. We drink coffee or tea, and the coffee grounds and tea leaves are filtered through nonwoven textiles. We dress in knit and woven apparel fabrics. When we get into a car or bus, we sit on upholstered seats and the vehicle moves on tires reinforced with strong textile yarns. We stand on carpets, sit on upholstered furniture, and look out of curtained(draped) windows in our living and working spaces. Fiberglass insulation in our buildings reduces heating and cooling bills. The golf clubs, tennis rackets, and ski poles we use in recreational sports may be reinforced with lightweight textile fibers. The roads and bridges we travel over may be stabilized or reinforced with textiles. The stadiums we sit in may be covered with a fabric roof. Some of us have wounded tissue closed and held together by textile sutures after surgical procedures. Throughout the day, we use other types of manufactured products; most of them would have more or less fibers or textiles embedded. Even the processed foods we eat have passed through textile filters. Our increased knowledge of space is partially due to the development of strong, heat-resistant textile fibers that prevent the exhaust cones of space satellites from melting and disintegrating under the tremendous heat they are exposed to when launched. The spacewalking astronauts must be protected by space suits against high temperature, solar radiation and micro-environmental factors such as meteor harm the human body in space. Nowadays, every person in home capable of knowing his or her health condition without complex detecting instruments is due to the use of smart fabrics or wearable devices. No aspect of our lives seems untouched by textiles.

The History of Textiles

The history of textiles is almost as old as that of human civilization. The wearing of clothing is exclusively a human characteristic and is a feature of most human societies.

It is not known when humans began wearing clothes because nature animal hair and plant fibers used were easily rotten but anthropologists believe that animal skins and vegetation were adapted into coverings as protection from cold, heat and rain, especially as humans migrated to new climates. Our knowledge of ancient textiles and clothing has expanded in the recent past thanks to modern technological developments. Genetic analysis suggests that the human body louse, which lives in clothing, may only have diverged from the head louse some 170 millennia ago, which supports evidence that humans began wearing clothing at around this time. These estimates predate the first known human exodus from Africa, although other hominid species who may have worn clothes-and shared these louse infestations-appear to have migrated earlier. Evidence suggests that humans may have begun wearing clothing as far back as 10, 000 to 50, 000 years ago. The earliest dyed flax fibers have been found in a prehistoric cave in the Republic of Georgia and date back to 36, 000 BP. Possible sewing needles have been dated to around 40, 000 years ago. More evidence would have to rely on the plethora of stunning cave paintings and rock drawings in prehistoric times.

Clothing and textiles have been important in human history and reflect the materials available to a civilization as well as the technologies that had been mastered. The social significance of the finished product reflects their culture.

Our knowledge of cultures varies greatly with the climatic conditions to which archeological deposits are exposed; the Middle East and the arid fringes of China have provided many very early samples in good condition, but the early development of textiles in other moist parts of the world remains unclear.

Textiles have a long tradition in the world. Historical evidence indicates that in the early days of human civilization, three principal countries of the east have been distinguished by their distinctive type of textiles: China as the land of silk, Egypt as the country of flax (and linen), and India of cotton. Let's just use Chinese textiles as an example.

Chinese textiles are world-famous and extraordinary for their fine quality and profound symbolic meanings. Chinese textiles are one of the oldest in the world, dating back to the Chinese civilization. The earliest evidence of silk production in China was found at the sites of Yangshao culture in Xia, Shanxi, where a cocoon of bombyx mori, the domesticated silkworm, cut in half by a sharp knife is dated to between 5000 BC and 3000 BC. Fragments of primitive looms are also seen from the sites of Hemudu culture in Yuyao, Zhejiang, dating back to about 4000 BC. Scraps of silk were found in a Liangzhu culture site at Qianshanyang in Huzhou, Zhejiang, dating back to 2700 BC. Other fragments have been recovered from royal tombs in the Shang Dynasty (c. 1600 BC—c.1046 BC).

Under the Shang Dynasty, Han Chinese clothing or Hanfu consisted of a yi, a narrow-cuffed, knee-length tunic tied with a sash, and a narrow, ankle-length skirt, called shang, worn with a bixi, a length of fabric that reached the knees. Clothing of the elite was made of silk in vivid primary colours. During the history of China, lots of artists presented the silk clothing in their work. For example, Daolian Drawing as shown in Fig. 1.2, one of famous paintings in Tang Dynasty, presents silk fabric manufacture in China.

思政微课堂：
捣练图的故事

Fig.1.2 Court ladies preparing newly woven silk (a part of Daolian Drawing collected in Museum of Fine Arts Boston in USA). Early 12th-century painting by Emperor Huizong of Song (a remake of an 8th-century original by artist Zhang Xuan in Tang Dynasty)

思政微课堂：
丝绸之路

With the growing of world's population, the rise of productivity, the increasing wealth of goods, it changes our ways of living. There is a road, Silk Road, that changed the world. The Silk Road is named after the lucrative international trade in Chinese silk textiles that started during the Han Dynasty (207 BC—220 CE), although earlier trade across the continents had already existed.

The exchange of luxury textiles was predominant on the Silk Road, a series of ancient trade and cultural transmission routes that were central to cultural interaction through regions of the Asian continent connecting East and West by linking traders, merchants, pilgrims, monks, soldiers, nomads and urban dwellers from China to the Mediterranean Sea during various periods of time. Geographically, the Silk Road or Silk Route is an interconnected series of ancient trade routes between Chang'an (today's Xi'an) in China, with Asia Minor and the Mediterranean extending over 8,000km on land and sea. Trade on the Silk Road was a significant factor in the development of the great civilizations of China, Egypt, Mesopotamia, Persia, the Indian subcontinent, and Rome, and helped to lay the foundations for the modern world.

Nowadays, the China-proposed Belt and Road Initiative (BRI) will help the mechanism boost trade among nations in the region and increase global real income.

The Industrial Revolution, began in Britain, was a major turning point in world history; Almost every aspect of daily life changed in some way.

The Industrial Revolution was the transition to new manufacturing processes in the period from about 1760 to sometime between 1820 and 1840.

During this transition, hand production methods changed to machines and new chemical manufacturing and iron production processes were introduced. Water power efficiency improved and the increasing use of steam power increased. Machine tools were developed and the factory system was on the rise. Average income and population began to grow exponentially. Textiles were the main industry of the Industrial Revolution as far as employment, the value of output and capital invested. The textile industry was also the first to use modern production methods.

Several inventions in textile machinery occurred in a relatively short time period during the Industrial Revolution. Here is a timeline highlighting some of them:

- 1733 Flying shuttle invented: an improvement to looms that enabled weavers to weave faster.
- 1742 Cotton mills were first opened in England.
- 1764 Spinning jenny invented: the first machine to improve upon the spinning wheel.
- 1764 Water frame invented: the first powered textile machine.
- 1773 The first all-cotton textiles were produced in factories.
- 1779 The spinning mule invented: greater control over the weaving process.
- 1787 Cotton goods production had increased 10 fold since 1770.
- 1790 The first steam-powered textile factory was built.

思政微课堂：
纺织工业

- 1804 The Jacquard loom invented: weaved complex designs.
- 1813 The variable speed loom invented.
- 1856 The first synthetic dye invented.

Categories of Textile Fibers

A textile fiber is a unit of matter, either natural or man-made, that forms the basic element of fabrics and other textile structures. Generally, textile fibers are used successfully for manufacturing three types of fabrics which are apparel fabrics, home furnishing fabrics and industrial fabrics of certain types. The major characteristic of textile fibers is thousand times longer than its width. Textile fibers generally fall within the range of 10 to 50μm in diameter and individual fibers can vary in length from less than 1cm to thousands of meters.

Textile fabrics are rarely manufactured from individual fibers but rather from a yarn, which is defined as a strand of textile fibers in a form suitable for wearing, knitting, braiding, felting, webbing, or otherwise fabricating into a fabric.

To become a textile fabric, fibers used must have some fundamental properties and characteristics, as given below:

(1) It must have fibrous formation.

(2) Its length is thousand times longer than its diameter.

(3) It should have sufficient strength and spinnability.

(4) It should have contained elasticity and flexibility characteristics.

(5) It must be fineness.

(6) It must have special color.

(7) It should have affinity to dye stuff.

According to the properties and characteristics, textile fibers are classified into two main parts which are natural fiber and man-made fiber or artificial fiber. The general classification of textile fibers is listed in Fig.1.3 as below.

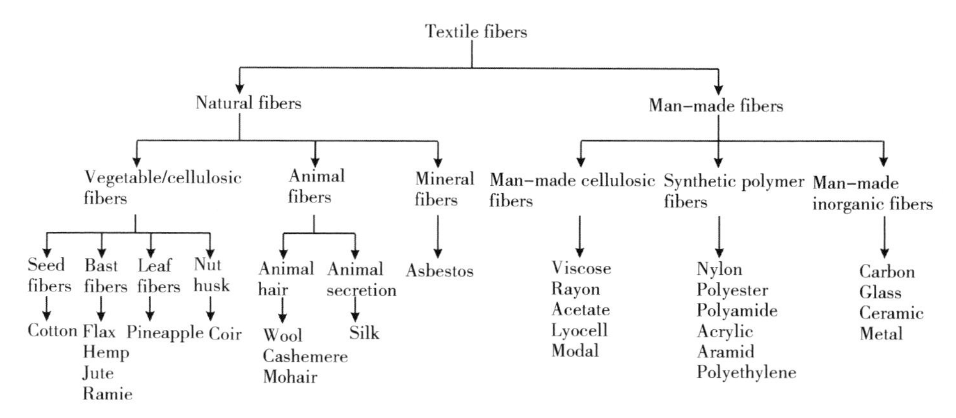

Fig.1.3 The general classification of textile fibers

A natural fiber is any fiber that exists as such in the natural state, such as cellulose, protein and mineral. The origin of cellulose is a vegetable or plant, protein is animal and mineral is asbestos. A man-made fiber is an artificial fiber, which is made by a polymerization process in the factory. But some man-made fibers are made from natural fibers which are called man-made cellulosic fiber or regenerated man-made fiber. Other man-made fibers are classified in two ways: synthetic polymer fibers and man-made inorganic fibers. Synthetic polymer fibers are formed from substances that, at any point in the manufacturing process, are not a fiber; examples are nylon and polyester. No nylon or polyester fibers exists in nature and they are made of chemicals put through reactions to produce the fiber-forming substance.

World fiber production equalled approximately 110 million metric tons in 2018, including 32 million tons of natural fibers and 79 million tons of chemical fibers, both cellulosic and non-cellulosic (from petroleum molecules). Cotton accounted for 80% of natural fiber production by weight, while polyester, particularly polyester filament, dominated the chemical fiber universe. Jute production totalled about 3 million tons in 2018, while wool and coir each accounted for about 1 million tons. Specialized natural fibers added another 1.5 million tons. The man-made fiber complex accounted for 70% of total fiber production in 2018. Since 1998, China's chemical fiber production has ranked first in the world. During 2011—2019, China's chemical fiber production accounted for over 70% of the world's total output that reported data every year.

New Words and Expressions

textile 纺织,纺织物,纺织品
fiber 纤维
articial fibers 人造纤维
originated 起源
weave 机织
knit 针织
braid 编织
nonwovens 非织,非织造布
spindle 纺锤,锭子
distaff 卷线杆
vegetable fiber 植物纤维
handloom 手工织布机
spinning machine 纺纱机
loom 织机
yarn 纱
thread 线
cloth 布,织物
spin 纺纱
cotton 棉
synthetic fiber 合成纤维
fabric 织物,布,面料
sheet 床单,被单
towel 毛巾,手巾
filter 过滤,过滤器
apparel 服装;衣服
reinforce 加强,加固
fiberglass 玻璃纤维,玻璃丝
lightweight 轻质的
heat-resistant 耐热的
space suits 宇航服,太空服
human civilization 人类文明
wear 穿
plant fibers 植物纤维
anthropologist 人类学家
louse 虱子
millennia 千年

predate 早于
dyed 染色的
flax 亚麻
prehistoric 史前的
Georgia 格鲁吉亚
sew 缝纫，缝
needle 针
wool 羊毛
silk 蚕丝
Yangshao culture 仰韶文化
cocoon 蚕茧
bombyx mori 家蚕
Hemudu culture 河姆渡文化
Liangzhu culture site 良渚文化遗址
Han Chinese clothing 汉服
Hanfu 汉服
Daolian Drawing 捣练图
Silk Road 丝绸之路
Asia Minor 小亚细亚
Industrial Revolution 工业革命
exponentially 指数级地
shuttle 梭子
spinning jenny 珍妮纺纱机
water frame 水力纺纱机
spinning mule 走锭纺纱机
Jacquard loom 提花织机
synthetic dye 合成染料
asbestos 石棉
chemical fiber 化学纤维

Notes to the Text

1. Latin word：拉丁语。
2. originated from：来源于，起源于。
3. genetic analysis：遗传分析，基因分析。
4. the Republic of Georgia：格鲁吉亚共和国。
5. cave paintings：洞穴画，洞穴壁画。
6. rock drawings：原始岩画，古代岩画。
7. Belt and Road Initiative：一带一路倡议。
8. polymerization process：聚合过程。
9. smart materials and wearable electronics allow textiles to become active and reactive：智能材料和可穿戴电子产品使纺织品变得主动和可交互。

Questions to the Text

1. Describe the relationship between modern life and textiles.
2. Describe the history of textiles.
3. Write a paper in at least 500 words to say something about textiles.
4. Describe the categories of textile fibers.
5. What are basic properties and characteristics of textile fiber?
6. Share your thoughts on the future trends of textiles with your classmates.

Lesson Two

Cotton

语言微课堂

Cotton is the world's most widely used natural fiber. Its popularity stems from both its relative ease of production and its applicability to a wide variety of textile products.

Before yarn manufacture, cotton is graded, sorted, and blended to insure uniform yarn quality. Cotton is graded on the basis of color, staple length, fineness, and freedom from foreign matter. In the United States, cottons are divided into grades according to length of staple, uniformity, strength, color, cleanness and flexibility. Theses are compared with a standard supplied by the United States Department of Agriculture. The standard provides 6 grades above and 6 grades below the Middling grade. The most common grades are:

(1) Strict good middling.

(2) Good middling.

(3) Strict middling.

(4) Middling.

(5) Strict low middling.

(6) Low middling.

(7) Strict good ordinary.

专业微课堂：棉

The cotton fiber may be from 0.3 to 5.5cm long. Under the microscope is appears as a ribbon like structure that is twisted at irregular intervals along its length [Fig.2.1(a)]. The twists, called convolutions, increase the fiber-to-fiber friction necessary to secure a strong spun yarn. The fiber ranges in color from a yellowish to pure white, and may be very lustrous. However, most cotton is dull.

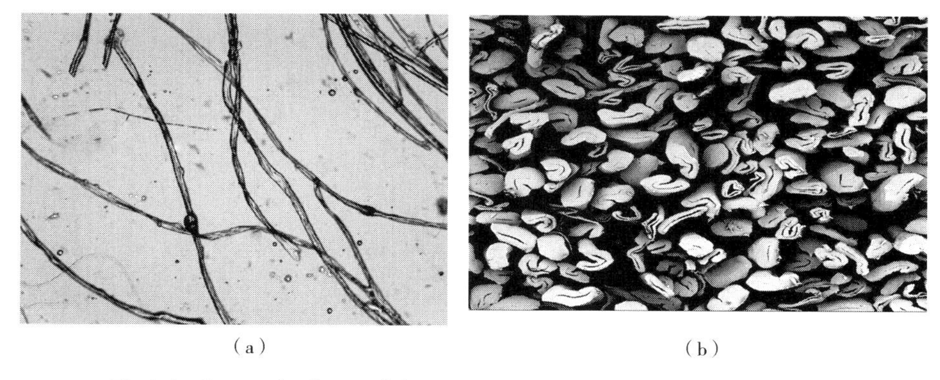

(a) (b)

Fig.2.1 Longitudinal view (a) and cross-section view (b) of cotton fibers

A cross-sectional view reveals that the fiber is kidney-shaped with central hollow core known as the lumen [Fig.2.1(b)]. The lumen provides a channel for nutrients while the plant is growing. The fiber consists of an outer shell, or cuticle, which surrounds the primary wall. The primary wall, in turn, covers the secondary wall surrounding the lumen. The cuticle is a thin, hard shell which protects the fiber from bruising and damage during growth. In use as a textile fiber, the cuticle provides abrasion resistance to cotton.

A relatively high level of moisture absorption and good wicking properties help make cotton one of the more comfortable fibers. Because of the hydroxyl groups in the cellulose, cotton has a high attraction for water. As water enters the fiber, cotton swells and its cross section becomes more rounded. The high affinity for moisture and the ability to swell when wet allow cotton to absorb about one-fourth of its weight in water. This means that in hot weather perspiration from the body will be absorbed in cotton fabrics, transported along the yarns to the outer surface of the cloth and evaporated into the air. Thus, the body will be aided in maintaining its temperature.

Unfortunately, the hydrophilic nature of cotton makes it susceptible to water-borne stains. Water-soluble colorants such as those in coffee or grape juice will penetrate the fiber along with the water; when the water evaporates, the colorant is trapped in the fiber. Perhaps the major disadvantage to cotton goods is their tendency to wrinkle and the difficulty of removing wrinkles. The rigidity of cotton fibers reduces the ability of yarns to resist wrinkling. When the fibers are bent to a new configuration, the hydrogen bonds which hold the cellulose chains together are ruptured and the molecules slide in order to minimize the stress within the fiber. The hydrogen bonds re-form in the new positions, so that when the crushing force is removed the fibers stay in the new positions. It is the rupture and reformation of the hydrogen bonds that helps to maintain wrinkles, so that cotton goods must be ironed.

Cotton is a moderately strong fiber with good abrasion resistance and good dimensional stability. It is resistant to the acids, alkalis and organic solvents normally available to consumers. But since it is a natural material, it is subject to attack by insects, molds and fungi. Most prominent is the tendency for cotton to mildew if allowed to remain damp.

Cotton resists sunlight and heat well, although direct exposure to constant strong sunlight will cause yellowing and eventual degradation of the fiber. Yellowing may also occur when cotton goods are dried in gas dryers. The color change is the result of a chemical reaction between cellulose and oxygen or nitrogen oxides in the hot air in the dryer. Cottons will retain their whiteness longer when line-dried or dried in the electric dryer.

Of major interest is the fact that cotton yarn is stronger when wet than when dry. This property is a consequence of the macro-and micro-structural features of the fiber. As water is absorbed, the fiber swells and its cross section becomes more rounded. Usually the absorption of such a large amount of foreign material would cause a high degree of internal stress and lead to weakening of the fiber. In cotton, however, the absorption of water causes a decrease in the internal stresses. Thus, with less internal stresses to overcome, the swollen fiber becomes stronger. At the same time, the swollen fibers within the yarns press upon each other more strongly. The internal friction strengthens the yarns. In addition, the absorbed water acts as an internal lubricant which imparts a higher level of flexibility to the fibers. This accounts for the fact that cotton garments are more easily ironed when damp. Cotton fabrics are susceptible to shrinkage upon laundering.

Perhaps more than any other fiber, cotton satisfies the requirements of apparel, home furnishings, recreational, and industrial uses. It provides fabrics that are strong, lightweight, pliable, easily dried, and readily laundered.

In apparel, cotton provides garments that are comfortable, readily dried in bright, long-lasting colors, and easy to care for. The major drawbacks are a propensity for cotton yarns to shrink and for cotton cloth to wrinkle. Shrinkage may be controlled by the application of shrink-resistant finishes. Durable-press properties may be imparted by chemical treatment or by blending cotton with more wrinkle-resistant fibers, such as polyester.

In home furnishings, cotton fabrics have been the mainstay of bed linens and towels for decades, because they are comfortable, durable, and moisture-absorbent. Polyester cotton blends provide the modern consumer with no-iron sheets and pillowcases that retain a crisp, fresh feel.

For recreational use, cotton has traditionally been used for tenting and camping gear, boat sails, tennis shoes and sportswear. Cotton is particularly well-suited for tent. Fabrics woven from cotton can be open enough to provide good air permeability for comfort. Tents should also shed water, when wet by rain, cotton yarns swell, reducing the interstices between the yarns and resisting the penetration of water.

In China, Xinjiang cotton has good properties and excellent quality. Xinjiang has abundant land resources, and its climate has low precipitation, high levels of sunshine and large temperature difference between day and night. Benefiting from these unique advantages, Xinjiang cotton shows good fiber length and fineness, good color and strength, and less foreign fiber content. Since the 1980s, Xinjiang Uygur Autonomous Region has gradually become one of the most important high-quality and high-yield cotton regions in both China and the world.

思政微课堂:新疆棉

New Words and Expressions

topographical 地形的,地貌的
foreign fiber 异纤
staple 短纤维
convolution 扭曲
spun yarn 短纤维纱线
lumen 腔
cuticle 表皮
primary wall 初生胞壁
secondary wall 次生胞壁
wicking 芯吸
hydroxyl groups 羟基
cellulose 纤维素
hydrophilic 亲水性的
susceptible 敏感的
water-borne stains 水基污渍
colorant 颜料,染料
hydrogen bond 氢键
stress 应力

iron 熨烫
moderately 中等的
organic solvent 有机溶剂
mold 霉菌
fungus 真菌
prominent 突出的，显著的
mildew 发霉
degradation 降解
gas dryer 煤气烘干机
impart 给予
flexibility 挠性
home furnishings 家庭装饰
recreational 休闲的
pliable 柔顺的
garment 服装
shrink-resistant finish 防缩整理
blending 混纺
polyester 涤纶
mainstay 主要力量
no-iron 免烫
pillowcase 枕套
crisp 挺爽
interstice 空隙
Xinjiang Cotton 新疆棉

Notes to the Text

1. cotton is graded, sorted：棉花被分等、分类。
2. freedom from foreign matter：含杂程度。
3. United States Department of Agriculture：美国农业部，在美国棉花种植与纺织工业由农业部负责。
4. 美国常用的棉花等级名称：
   Strict good middling 次优级棉(二级)
   Good middling 上级棉(三级)
   Strict middling 次上级棉(四级)
   Middling 中级棉(五级)
   Strict low middling 次中级棉(六级)
   Low middling 下级棉 (七级)
   Strict good ordinary 次下级棉(八级)
5. cotton goods：棉织物，goods 通常可表示织物。
6. It is resistant to the acids, alkalis, and organic solvents normally available to consumers：句子中"normally available to consumers" 可以译为"家庭使用的"。
7. it is subject to attack by insects：句子中"subject" 意思为 "易受……伤害的"。
8. line-dried：挂在绳子上自然晾干。
9. This accounts for the fact that：这就是……的原因。
10. Cotton fabrics are susceptible to shrinkage upon laundering：句子中"are susceptible to" 意思为"容易的，易受影响的"。
11. durable-press：耐久压烫，耐久定形整理(简称 DP 整理)。
12. bed linen：床单(由于早期的床单由亚麻制作，故得此名。在当代，尽管制作床单的原料并不只限于亚麻，但人们依然习惯用"bed linen"表示)。
13. tenting and camping gear：帐篷(gear 的意思是"用具，装具")。

14. Fabrics woven from cotton can be open enough to：句子中的“open”是指织物的稀疏，“open fabric”表示纱线密度极小的织物。
15. Xinjiang Uygur Autonomous Region：新疆维吾尔自治区。

Questions to the Text

1. What are the main considerations in working out the cotton grades in the United States?
2. Describe the shape character of the cotton fiber.
3. What can we benefit from the good moisture absorption of the cotton?
4. Why cotton goods are easier to be stained by water-borne materials?
5. Describe the wrinkle generating process.
6. Cotton yarn is stronger when wet than dry, why?
7. Say something about the usages of the cotton fiber.
8. Write a paper in at least 500 words to say something about the cotton.

语言微课堂

Lesson Three

Properties of the Naturally Colored Cottons *

Textile is among the industries in which harmful substances would arise. The disposal of the drainage after dyeing and finishing causes environmental pollution. On the other hand, questions are frequently raised regarding textile fibers and their impact on the environment, workplace and their product safety. It is reported that fabrics made by using synthetic fibers would cause health hazards to the users.

Cotton, here means the white cotton, has been the most important textile material. It is understood that bleaching, mercerizing and dyeing of the cotton goods would cause environmental problems. The usage of pesticide, chemical fertilizer, and weed killer during cotton planting, the chemical treatment during textile manufacturing, lead to chemical deposits in the cotton fiber, these are harmful to the customers.

Naturally colored cotton is considered to be disease resistant during its growth, no traditional pesticides are needed. If organic fertilizer is employed, the grown colored cotton may be regarded as totally chemical free. Naturally colored cotton is environmentally friendly as its goods need no dyeing. Colored cotton is not a new concept, its cultivating history can date back to 5,000 years ago. Later, owing to its lower yield and inferior quality compared with that of the white cotton, colored cotton had not been frequently mentioned until the 1970s, when people began to aware the importance of the environmental protection. Breeding research over the last 30 years in this field has led to improvement in yields, fiber quality and color intensity. China began its research work on colored cotton in the 1990s. To date, 4 colored cotton research centers have been set up, plantations specialized in colored cotton have been run successfully for continuous several years in the provinces of Sichuan, Gansu, Henan, Xinjiang and Hainan. Colors of the cultivated cotton include brown, green, purple, grey and orange. It is estimated that in 30 years the colored cotton and organic cotton would take 30% of its total cotton output of the world.

In order to use the colored cotton effectively, thorough understanding of the colored cotton fibers is necessary.

专业微课堂：
棉花的一生

思政微课堂：
彩色棉

Structures and properties of the colored cottons

Table 3.1 gives some of the substances extracted from the colored cottons. White cotton is included for comparison.

Table 3.1 Some compositions of the colored cotton

Substance	White cotton	Brown cotton	Green cotton
Fat/%	0.6	3.19	4.34
Lignin/%	0	638	9.34
Pectin/%	1.2	0.43	0.51

The table reveals that the fat content in the colored cottons is larger than that of the white cotton. And there is no lignin in the white cotton when larger amount of it is noticed in the colored cottons. As has been recognized that both fat and lignin are considered to be hydrophobic. One may estimate that the ability of the moisture absorption of the colored cotton goods would be lower than that of the white. The table also shows that the pectin content in the white cotton is the highest.

Table 3.2 illustrates the major physical properties of the colored cotton fibers. For the sake of comparison, the white cotton is also included.

Table 3.2 Cotton fibers physical properties

Item	White cotton	Brown cotton	Green cotton
Fineness/(m/g)	5, 700	5, 930	6, 750
Length/mm	28.59	28.76	26.35
Tenacity/(g/denier)	4.03	3.98	2.45

Usually the colored fibers are finer than the white fiber, especially for the green one, it measures 6, 750m/g. The strength of the finer fibers is usually weaker than that of the thicker ones, anyway, the yarn evenness may be better because more fibers can be included in a definite yarn cross section.

The brown cotton fiber is the longest, when fibers are twisted to form the yarn, longer fibers offer more connection areas among the fibers, resulting in greater friction among the fibers and hence the yarn strength. Compared to other fibers, the tenacity of the green cotton fiber is the lowest. One may expect its lower yarn strength.

Suggestions in using the colored cottons

Based on the physical properties of the colored cottons and the spinning practice, some suggestions are listed below for using the colored cottons effectively and economically.

(1) Naturally colored cottons are recommended to blend with white cotton because they are usually shorter, finer and weaker and because of their higher price. Add certain percentage of white cotton in the yarn can reduce the initial cost of the yarn. On the other hand, the color shade of the yarn can be adjusted by using various blending ratios.

(2) Blending with synthetic fibers. Synthetic staples such as polyester, nylon etc. can be blended with colored cotton in yarn spinning. It is recommended that the content of the synthetic fiber should not be higher than 10%. This can increase the strength of the yarn greatly. Too much synthetic fiber content will deteriorate the environmental character of the colored cotton.

(3) Since the limited color variety, colored cottons with different hue can be blended to spin the yarn of expected colors.

New Words and Expressions

disposal 处理，排放
drainage 污水
mercerize 丝光处理
pesticide 杀虫剂
weed killer 除草剂
deposit 存留，囤积
organic fertilizer 有机肥
environmentally friendly 环境友好地
breed 育种，品种
color intensity 颜色强度
plantation 农场
lignin 木质素
pectin 果胶
hydrophobic 疏水的
deteriorate 损害
hue 色彩

Notes to the Text

1. On the other hand, questions are frequently raised regarding textile fibers and their impact on the environment：句中“their impact on the environment”指它们对环境的影响。
2. the yarn evenness may be better because more fibers can be included in a definite yarn cross section：在一定细度的纱线截面内可以包含较多的纤维，纱线的均匀度较好。
3. resulting in greater friction among the fibers and hence the yarn strength：句中“and hence the yarn strength”因此导致强度较大，其中省略了“greater”。
4. 1 g/denier = 1 g/d = 1 gf/den ≈ 0.0882 N/tex ≈ 8.82 cN/tex.

Questions to the Text

1. Why the colored cottons were forgotten for a period in the history?
2. Why some textiles are harmful to consumers?
3. Compare the physical properties of the cottons in different color.
4. What are the suggestions of the author when using the colored cotton?

语言微课堂

Lesson Four

Wool

The early history of wool is lost in antiquity. Sheepskin, including the hair, was probably used long before it was discovered that the fibers could be spun into yarns or even felted into fabric. There is no evidence to support the theory that wool was the first fiber to be processed into fabric, but it seems certain that, as a part of the skin, wool was used for covering and protection by prehistoric people long before yarns and fabrics were made.

The earliest fragments of wool fabric have been found in Egypt, probably because of the preserving qualities of the climate. These have been dated from 4, 000 to 3, 500 BC. The earliest example of wool fabric found in Europe has been dated about 1, 500 BC. It was unearthed in archeological digs in Germany. Danish sites have yielded excellent fragments of early wool fabrics dated about 1, 300 to 1, 000 BC. These fabrics are rough and coarse and contain considerable wild sheep hair.

Under the microscopic observation, the length of the wool fiber shows a scale structure. The schematic diagram of the structure of Merino wool is shown in Fig.4.1. The size of the scale varies from very small to comparatively broad and large. As many as 700 scales are found in 1cm of fine wool, whereas coarse wool may have as few as 275 per cm. Fine wool does not have as clear and distinct scales as coarse wool, but they can be identified under high magnification.

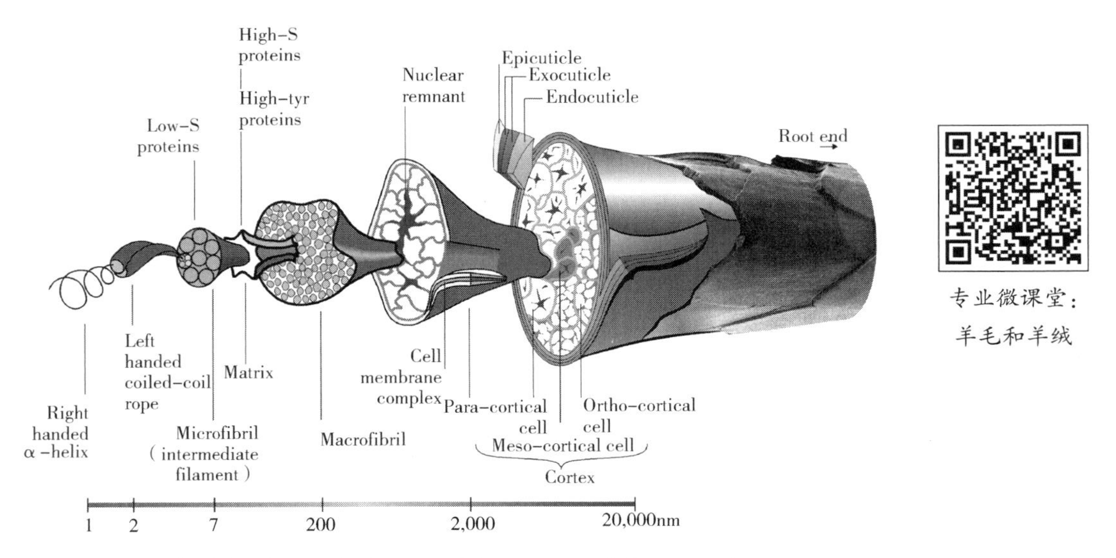

Fig.4.1 Schematic diagram of the structure of Merino wool

A cross section of wool shows three distinct parts to the fiber. The outer layer, called cuticle, is composed of the scales. These scales are somewhat horny and irregular in shape, and they overlap, with the top pointing towards the tip of the fiber; they are similar to fish scales. The major portion of the fiber is the cortex (composed of cortical cells); this extends toward the center from the cuticle layer. Cortical cells are long and spindle-shaped and provide fiber strength and elasticity. The cortex accounts for approximately 90% of the fiber mass. In the center of the fiber is the medulla. The size of the medulla varies and in fine fibers may be invisible. This is the area through which food reached the fiber during growth, and it contains pigment that gives color to fibers.

Wool fibers vary in length from 3.8 to about 38cm. Most authorities have determined that fine wools are usually from 3.8 to 12.7cm; medium wool from 6.4 to 15.2cm; and long (coarse) wools from 12.7 to 38cm.

The width of wool also varies considerably. Fine fibers such as Merino have an average width of about 15 to 17 microns; whereas medium wool averages 24 to 34 microns and coarse wool about 40 microns. Some wool fibers are exceptionally stiff and coarse; these are called kemp and average about 70 microns in diameter.

The wool fiber cross section may be nearly circular, but most wool fibers tend to be slightly elliptical or oval in shape. Wool fibers have a natural crimp, a built-in waviness. The crimp increases the elasticity and elongation properties of the fiber and also aids in yarn manufacturing. It is three-dimensional in character; in other words, it not only moves above and below a central axis but also moves to the right and left of the axis.

There is some luster to wool fibers. Fine and medium wool tends to have more luster than very coarse fibers. Fibers with a high degree of luster are silky in appearance.

The color of the natural wool fiber depends on the breed of sheep. Most wool, after scouring, is a yellowish-white or ivory color. Some fibers may be gray, black, tan or brown.

The tenacity of wool is 1.0 to 1.7 grams per denier when dry; when wet, it drops to 0.7 to 1.5 grams per denier. Compared with many other fibers, wool is weak, and this weakness restricts the kinds of yarns and fabric constructions that can be used satisfactorily. However, if yarns and fabrics of optimum weight and type are produced, the end-use product will give commendable wear and retain shape and appearance. Fiber properties such as resiliency, elongation and elastic recovery compensate for the low strength.

Wool has excellent elasticity and extensibility. At standard conditions the fiber will extend between 20 and 40%. It may extend more than 70% when wet. Recovery is superior. After a 2% elongation the fiber has an immediate regain or recovery of 99%. Even at 10% extension, it has a recovery of well over 50%, which is higher than for any other fiber except nylon.

The resiliency of wool is exceptionally good. It will readily spring back into shape after crushing or creasing. However, through the application of heat, moisture and pressure, durable creases or pleats can be put into wool fabrics. This crease or press retention is the result of molecular adjustment and the formation of new cross-linkages in the polymer. Besides resistance to crushing and wrinkling, the excellent resilience of wool fiber gives the fabric its loft, which produces open, porous fabrics with good covering power, or thick, warm fabrics that are light in weight. Wool is very flexible and pliable, so it combines ease of handling and comfort with good shape retention.

The standard moisture regain of wool is 13.6 to 16.0%. Under saturation conditions, wool will absorb more than 29% of its weight in moisture. This ability to absorb is responsible for the comfort of wool in humid, cold atmospheres. As part of the moisture absorption function, wool produces or liberates heat. However, as wet wool begins to dry, the evaporation causes heat to be absorbed by the fiber, and "chilling" may be experienced, though the chilling factor is slowed down as the evaporation rate is reduced. The property of moisture absorption and desorption peculiar to wool and similar hair fibers is called hygroscopic behavior. Wool accepts colors and finishes easily because of its moisture absorption properties.

Despite the absorption properties of wool, it has an unusual property of exhibiting hydrophobic characteristics. That is, it tends to shed liquid easily and appears not to absorb moisture. The cause is a combination of factors: interfacial surface tension, uniform distribution of pores, and low bulk density. These moisture properties help make wool very desirable for use in a variety of situations where moisture can be a problem to comfort.

Wool fibers are not dimensionally stable. The structure of the fiber contributes to a shrinking and felting reaction during processing, use and care. This is due, in part, to the scale structure of the fiber. When subjected to heat, moisture, and agitation, the scales tend to pull together and move toward the fiber tip. This property is noticeable in yarns and fabrics and is responsible for both felting and relaxation shrinkage.

思政微课堂:
张汉文教授

思政微课堂:
羊绒发展历程

New Words and Expressions

scale 鳞片
cuticle 角质层
horny 角状的
cortex 角质，皮质
cortical cell 角质细胞，皮质细胞
medulla 毛髓
kemp 死毛，戗毛
silky 像蚕丝一样的
durable crease 耐久皱褶
pleat 皱褶，打褶
cross-linkage 交键
polymer 聚合物
loft 蓬松的，高雅的
covering power 覆盖系数

moisture regain 回潮率
desorption 水分释放
hygroscopic 吸湿的
finish 织物的整理
interfacial 界面的
felting 毡化

Notes to the Text

1. a built-in waviness：内在的卷曲，“built-in”意思为“内在的，内部具有的”。
2. it not only moves above and below a central axis but also moves to the right and left of the axis：它不但能绕轴线上下移动，也可以绕轴线左右移动（表示羊毛纤维是一种三维空间结构）。
3. resiliency：回弹性；该词与“elongation recovery”有区别。“resiliency”是指纤维受到挤压后变形的回复能力，而“elongation recovery”是指纤维受到拉伸后变形的回复能力。
4. so it combines ease of handling and comfort with good shape retention：因此羊毛容易处理（加工），舒适性好，具有良好的形状稳定性（保型性）。

Questions to the Text

1. Describe the number of scales on the wool fiber。
2. Describe the structure characteristics of wool like length, width, cross section, shape etc.
3. How is the elasticity of the wool fiber?
4. How is the water absorption ability of the wool fiber?
5. What are the reasons of felting of the wool fiber?

Lesson Five

Silk

语言微课堂

For approximately 3, 000 years China successfully held the secret of silk and sericulture and held a virtual monopoly on the silk industry. About A D 300 Japan learned the secret of raising silkworms and reeling the filaments from the cocoons.

Throughout history, silk has maintained a position of great prestige and is considered a luxury fiber. It is often called the"queen of fibers". Perhaps one of the most important contributions silk has made in the history of textiles is that it was responsible for investigation into the possible production of man-made fibers. Scientists observed how the silkworm spun the fibers and believed that people could duplicate the art.

Silk is one of the strongest natural fibers used in creating textile products. It has a tenacity of 2.4 to 5.1 grams per denier when dry. Wet strength is about 80 to 85% of the dry strength. Silk has good elasticity and moderate elongation, when it is dry, the elongation varies from 10 to 25%; when wet, silk will elongate as much as 33 to 35%. At 2% elongation the fiber has a 92% elastic recovery.

Silk has medium resiliency. Creases will hang out relatively well, but not so quickly or completely as for wool.

The density of specific gravity of silk is cited as 1.25 to 1.34 grams per cubic centimeter, depending on the resource used. Because of the nature of silk, it is possible that the density varies among fibers as well as between the various types of moths that form the fiber. Another source of variation may be due to methods used in determining density. In any case, the density results in the formation of lightweight but strong filaments, yarns and fabrics.

Silk has a relatively high standard moisture regain of 11%. At saturation the regain is 25 to 35%. This relatively high absorption is helpful in applying dyes and finishes to silk; however, unlike many fibers, silk also absorbs impurities such as metal salts. These contaminants tend to damage silk by weakening the fiber or causing actual ruptures to occur when the fabric is not handled properly.

专业微课堂：
丝绸的生产过程

思政微课堂：
丝绸

Silk will ignite and continue to burn when there is another source of flame. After removal from the source, it will sputter and eventually extinguish itself. It leaves a crisp, brittle ash and gives off an odor like that of burning hair or feathers. It burns similarly to wool.

Like other protein fibers, silk has a lower thermal or heat conductivity than cellulosic fibers. This factor, coupled with certain methods of construction, creates fabrics that tend to be warmer than comparable fabrics of cellulosic fibers.

Silk is damaged by strong alkalies and will dissolve in heated caustic soda (NaOH); however, silk reacts more slowly than wool, and frequently the identity of the two fibers can be determined by the speed of solubility in NaOH. Weak alkalies such as soap and ammonia cause little or no damage to silk unless they remain in contact with the fiber for a long time.

Silk protein, like wool, can be decomposed by strong mineral acids. Medium concentrations of hydrochloric acid will dissolve silk, and moderate concentrations of other mineral acids cause fiber contraction and shrinkage. The molecular arrangement in silk permits rapid absorption of acids but tends to hold the acid molecules, so they are difficult to remove. This accounts for some of the acid damage to fibroin that does not occur to keratin. Organic acids do not damage silk and are used in some finishing processes. Some authorities maintain that the scroop of silk—a rustling or crunching sound which used to be considered a natural characteristic, is actually developed by exposure to organic acids.

Silk has been the queen of fabrics for centuries. As in the past, it is still used for luxury fabrics and for high-fashion items. It is frequently considered to be a sensuous fabric because of its smooth and soft feel, or hand.

Dry cleaning is the preferred method of care for silk fabrics and products. If handled carefully, however, silk fabrics can be laundered. A mild soap or synthetic detergent in warm, not hot, water should be used, and minimal handling is recommended. Thorough rinsing is required, and the best method for extracting water is to roll the garment in a towel and then hang it in a cool place, out of the sun, to dry. Tumble drying should not be attempted unless a care label specifies that such procedures are acceptable. Silk should be ironed or pressed at medium to low temperatures; steam is acceptable.

When silk requires bleaching, hydrogen peroxide or perborate bleaches must be used, as chlorine bleaches may destroy the silk.

One problem with silk is that body perspiration tends to weaken the fibers and frequently will alter the color. Many deodorants and antiperspirants contain aluminum chloride, which damage silk. It is advisable to wear protective dress shields if perspiration is a serious problem.

Several factors are involved in the demand for silk. It offers an incredible variety in fabric and yarn structure. Through dyeing, many beautiful fabrics can be produced. Probably no other fiber is so widely accepted and suitable for various occasions. It is versatile and can be used in almost any type of apparel and in a wide variety of fabrics for home furnishings.

Many silk fabrics cost considerably more than similar fabrics for man-made fibers. However, the consumer who has formed an attachment to silk is willing to pay the high price. Because silk combines strength, flexibility, good moisture absorbency, softness, warmth, luxurious appearance, and durability, choice products for the discerning consumer are made of this fiber. Its use, however, is limited primarily to apparel and home furnishings such as draperies and accessories.

New Words and Expressions

sericulture　养蚕业，蚕丝业
silkworm　蚕
reeling　缫丝，络丝
cocoon　蚕茧
queen of fibers　纤维皇后
denier　旦，旦尼尔
cellulosic fiber　纤维素纤维
fibroin　丝心蛋白
keratin　角蛋白
scroop　丝鸣
rustling　沙沙响
hand　手感
synthetic detergent　合成洗涤剂
minimal handling　尽量缓和的操作
rinsing　清洗，漂清
tumble drying　转笼烘干
care label　使用说明标签
hydrogen peroxide　过氧化氢
perborate　过硼酸盐
chlorine　氯
deodorant　除臭剂
antiperspirant　除汗剂
aluminum chloride　氯化铝
dress shield　防汗衬布
choice product　精品
drapery　帷幕，悬挂织物

Notes to the Text

1. Perhaps one of the most important contributions silk has made in the history of textiles is that it was responsible for investigation into the possible production of man-made fibers：蚕丝在历史上重要的贡献是引起了对人造纤维生产可行性的研究。
2. Creases will hang out relatively well：“hang out”通过悬挂可以消除。
3. is cited：引用，可译为“一般认为”。
4. is helpful in applying dyes and finishes to silk：有利于染色与整理。
5. coupled with certain methods of construction：再加上结构的一定变化。
6. comparable fabrics：指纱线线密度、纱线密度、织物结构相似的织物。
7. tends to hold the acid molecules：会截获酸的分子。
8. account for：是……的原因。
9. some authorities maintain：一些权威机构认为。
10. has formed an attachment to silk：青睐丝织品。
11. Organic acids：有机酸类是分子结构中含有羧基(—COOH)的化合物。在中草药的叶、根、特别是果实中广泛分布，如乌梅、五味子、覆盆子等。常见的植物中的有机酸有脂肪族的一元、二元、多元羧酸，如酒石酸、草酸、苹果酸、枸橼酸、抗坏血酸(即维生素C)等，也有芳香族有机酸，如苯甲酸、水杨酸、咖啡酸等。除少数以游离状态存在外，一般都与钾、钠、钙等结合成盐，有些与生物碱类结合成盐。脂肪酸多与甘油结合成酯或与高级醇结合成蜡，有的有机酸是挥发油与树脂的组成成分。

Questions to the Text

1. Say something on the properties of the silk.
2. What would happen when silk meets alkalies and acids?
3. What is scroop?
4. What care should be taken when cleaning the silk dresses?
5. Why silk is widely welcomed by the consumers?

Lesson Six

Flax

语言微课堂

Flax is a bast fiber—a woody fiber obtained from the phloem plants. It derives from the stalk or stem of the plant. Flax seed is planted in April or May. When the crop is to be used for fiber, the seed is sown close together so that the plants will be closely packed and produce fine plants with long, thin stems. The plants grow to a height of 0.9 to 1.2 meters for fiber use. The blossoms are a delicate pale blue, white, or pink. Flax for fiber is pulled before the seeds are ripe.

思政微课堂：麻

Flax for fiber is pulled by hand in some countries or by mechanical pullers. It is important to keep the roots intact, as fibers extend below the ground surface. Harvesting occurs in late August when the plant is a brown color. After drying, the plant is ripped; that is, it is pulled through special threshing machines that remove the seed bolls or pods.

To obtain fibers from the stalk, the outer woody portion must be rotted away. This process, known as retting, can be accomplished by any of several procedures.

Dew retting involves the spreading of the flax on the ground, where it is exposed to the action of dew and sunlight. This natural method of retting gives uneven results but provides the strongest and most durable linen. It requires a period of 4 to 6 weeks.

Pool retting is a process whereby the flax is packed in sheaves and immersed in pools of stagnant water. Bacteria in the water rot away the outer stalk covering. The time required is 2 to 4 weeks.

Tank retting similar to pool retting, utilizes large tanks in which the flax is stacked. The tanks are filled with warm water, which increases the speed of bacterial action. Tank retting requires only a few days. Both pool and tank retting give good-quality flax that is uniform in strength and light in color.

Chemical retting is accomplished by stacking the flax in tanks, filling the tanks with water, and adding chemicals such as sodium hydroxide, sodium carbonate, or dilute sulfuric acid. Chemical retting can be completed in a matter of hours instead of days or weeks. However, it must be carefully controlled in order to prevent damage to the fiber.

After the retting is complete, the flax is rinsed and dried. The stalks are then bundled together and pressed between fluted rollers that break the outer woody covering into small particles. It is then scutched to separate the outer covering from the usable fiber.

After scutching, the flax fibers are hackled, or combed. This operation separates the short fibers, called tow, from the long fibers, called line. This is accomplished by drawing the fibers between several sets of pins, each successive set finer than the preceding set. This process is similar to the carding and combing operation used for cotton and prepares the flax fibers for the final steps in yarn manufacture. As the fibers are removed from the hackling machine, they are drawn out into a sliver.

The flax sliver is drawn out into yarn, and twist is imparted. Flax fibers are spun either dry or wet, but wet spinning is considered to give the best quality yarn. The final yarn processing is similar to that used for cotton fiber.

Flax fiber is not so fine as cotton; flax cells have an average diameter of 15 to 18μm and vary in length from 0.63 to 6.35cm. Bundles of cells form the actual fiber as it is used in spinning into yarns, and these bundles may be anywhere from 12.7 to 50.8cm long. Line fibers are usually more than 30.5cm long; tow fibers are shorter.

The natural color of flax varies from light to ivory to gray. The choice fibers from Belgium are a pale sandy color and require little or no bleaching.

Flax fibers have a high natural luster with an attractive sheen.

Flax is a strong fiber; the normal tenacity ranges between 5.5 and 6.5 grams per denier. Some fibers of inferior quality may have a tenacity as low as 2.6 grams per denier, and some top quality fiber may exhibit a tenacity as high as 7.7 grams per denier. Fabrics of flax are durable and easy to maintain because of the fiber strength. When wet, the fiber is about 20% stronger than when dry.

Most linen fabrics for apparel use have been given various resin finishes to provide consumers with easy-care performance. These finishes reduce the strength of flax so that these fabrics tend to give less durability.

Flax has low pliability or flexibility, which may result in reduced serviceability in uses where frequent bending is required.

Flax has a standard moisture regain of about 12%. The saturation regain is comparable to that of other cellulosic fibers. Flax has outstanding wicking properties, which make it possible to move moisture along the fibers and yarns as well as to absorb moisture.

Like other cellulose fibers, flax burns quickly. It is highly resistant to decomposition or degradation by dry heat and will withstand temperatures to 150℃ for long period with little or no change in properties. Prolonged exposure above 150℃ will result in gradual discoloration. Safe ironing temperature may go as high as 260℃ as long as the fabric is not held at that high temperature for any length of time.

The strength of flax fibers makes it possible to manufacture a wide variety of yarns, from very fine to very heavy, which can be used to make a wide variety of fabrics, from the sheer and loose to heavy and compact. Linen is a frequent choice for table coverings because it wears well, looks extremely attractive and elegant, and, when properly finished, lies flat on the table. To achieve this flat effect, a beetling finish is often used.

The natural resistance of flax to chemicals, including detergents, bleaches, other laundry aids, and dry-cleaning solvents, provides a fabric that is easily maintained. Further, these properties, plus resistance to sunlight, inherent fiber strength, and resistance to aging, result in fabrics with a long life.

When selecting apparel items of linen, consumers prefer fabrics with crease-resistant or durable-press finishes. These may require special care instructions, which should be carefully followed.

New Words and Expressions

bast fiber　韧皮纤维
woody　木质的
phloem　韧皮
stalk　茎
stem　茎
intact　未经触动的
rip　剥
boll　植物的铃
rot　腐烂
retting　沤麻
dew retting　露水沤麻
pool retting　池塘沤麻
sheaves (sheaf 的复数形式)　捆，束
stagnant　不流动的，污浊的
bacteria (bacterium 的复数形式)　细菌
tank retting　池浸沤麻
sodium hydroxide　氢氧化钠
sodium carbonate　碳酸钠
dilute sulfuric acid　稀硫酸
fluted roller　沟槽罗拉
scutch　打麻
hackle　梳麻
comb　梳理
tow　短纤维，短麻，落纤
line　长纤维，长麻
carding　梳理，粗梳
sliver　条子
choice fiber　上等纤维
linen　亚麻
sheen　光彩，光泽
resin finish　树脂整理
easy-care　免烫
serviceability　耐用性能
saturation regain　饱和回潮率
wicking property　芯吸性能
discoloration　脱色
table covering　桌布
laundry aids　洗涤剂
aging　老化
durable-press　耐久压烫

Notes to the Text

1. pale blue：发白的淡蓝色。
2. mechanical puller：剥麻机。
3. threshing machine：打麻机。
4. each successive set finer than the preceding set：后套的针比前套的针要细。
5. from light to ivory to gray：颜色从白色，到象牙色，到灰色。
6. pale sandy color：白沙色。
7. as long as the fabric is not held at that high temperature for any length of time：只要不在这种高温下滞留即可。
8. beetling finish 捶布整理(通过捶打织物，使纱线扁平而呈现光泽，是织物增光、柔软

的整理方法)。

9. because it wears well：在这里“wear”是指亚麻织物做桌布时非常贴合桌面。

Questions to the Text

1. How many methods can be used for retting?
2. Say something about the tenacity of the flax fiber.
3. Why linen is a frequent choice for table coverings?
4. Give a report about the items made of flax you know.

Lesson Seven

语言微课堂

Other Bast Fibers*

专业微课堂：
韧皮纤维

Jute

Jute is a bast fiber like flax. It has been used since the dawn of the civilization; however, it did not attain economic importance until the latter part of the eighteenth century. Today, jute is one of the most widely used fibers in the world and is of special interest to countries with low economic standard, as it is a low cost fiber. Jute is processed into fabrics in Asia, Scotland, and the United States.

Jute is obtained from the stem of the plant. The major growing areas are Bangladesh, India, and Thailand. In fabric form it is frequently called burlap or hessian.

The jute plant is cultivated in a manner similar to that used for flax. The seeds are planted close together and grow to a height of 4.6 to 6.1m. The fibers are extracted from the stem by retting followed by breaking and scutching.

Natural jute has a yellow to brown or grayish color and a silky luster. Fibers are irregular in diameter and vary from 1.5 to 3m in length. The microscopic appearance is similar to flax except that there is less lumen in jute than flax. Fibers used in yarn construction are really bundles of fibrous material held together by a gummy substance. It is difficult to bleach jute to a pure white; thus, many jute fabrics are naturally beige to brown in color.

The average strength of the jute fiber is about 3.5 grams per denier; it has low elastic recovery, only 74% at 1% extension, and a low elongation—less than 2%. The resiliency is low, and yarns and fabrics that have been crushed or creased do not return to shape without treatment such as laundering and ironing.

The density of jute is about 1.5 grams per cubic centimeter; moisture regain is 13.7% at standard conditions. Jute will deteriorate, however, if held in moist areas for extended periods of time. Dry jute will last for a long time and retain much of its strength during that period. Dimensional stability is good.

Chemical reactions of jute are much like those of other cellulosic fibers. Resistance to alkalis and dilute acids is good. Concentrated acids do cause fiber breakdown. Sunlight does not damage jute.

Reactions to heat and flame are the same as for other cellulosic fibers. Jute has good resistance to microorganisms in some end uses.

Major end uses of jute include bagging, carpet backing, and furniture construction. It is used for bagging because it permits stacking without slipping or shifting position. In the manufacture of furniture, jute is used as a base fabric and as taping before surface fabrics are applied. As a carpet backing, jute adds stability and reduces slippage. It also is used as a base for the manufacture of linoleum. The amount of jute used for carpets and linoleum is decreasing yearly as man-made fibers, particularly olefins, are taking the place of jute.

Care of jute is similar to that of flax except for the fact that colored jute may require dry cleaning. Jute is widely used in developing countries and others where labor is available at relatively low cost. It is a labor-intensive commodity and requires a high proportion of labor in processing.

Ramie

China grass, or ramie, is a bast fiber that has been cultivated for hundreds of years in China. In most recent times it has been of interest to many other countries, including the United States.

The ramie plant, a member of the nettle family, is a perennial shrub that can be cut several times a season after the necessary preliminary growth period. It can be started from seeds, which necessitates a development period of three years before fibers are formed; or it can be grown from root cuttings, which mature within two years.

After cutting, the ramie stalks are retted to remove the outer woody covering and reveal the fine fibers within. These are degummed in caustic soda to eliminate the pectins and waxes, then bleached and finally neutralized in a dilute acid bath. The fiber is then washed and dried.

Ramie fibers are long and very fine. They are white and lustrous and almost silklike in appearance. The strength of ramie is outstanding, from 5.3 to 7.4 grams per denier. Elastic recovery is low and elongation is poor. The fibers are somewhat brittle and stiff.

Ramie reacts chemically in the same manner as other cellulosic fibers. The high degree of molecular crystallinity and low molecular accessibility reduce the rate of acid penetration and thus increase the time required for damage to occur as a result of strong acids.

Of special interest is the fact that ramie is highly resistant to microorganisms and insects. This is probably due to the presence of non-fibrous matter, which may contain material that is toxic to bacteria and fungi.

The use of ramie is increasing because of increased production of the fiber in Asia, particularly in China. A fairly large amount of fabric in which ramie has been blended with silk, linen, cotton, and some manufactured fibers, is appearing on the market. Fabrics of 100% ramie appear to be popular for apparel and some home furnishings.

Hemp

Hemp is a bast fiber that was probably used first in Asia. Records indicate that it was cultivated in China before 2300 BC. Latter it was carried into Europe and became an important fiber.

Today, hemp is grown on every continent and in nearly every country. A tall herb of the mulberry family, it is a tough plant and will grow at altitudes to 2,400m and in climates where temperatures are warm or hot. It can be replanted in the same fields without depleting soil nutrients.

The plants are cut by hand and then processed in a manner very much like that used for flax. It requires retting, followed by stripping or scutching to obtain the fiber. Then hackling, drawing, and spinning are needed to obtain the yarn. The fiber is dark tan or brown and is difficult to bleach; however, it can be dyed bright and dark colors successfully.

Hot concentrated alkalis will dissolve hemp, but hot or cold dilute or cold concentrated alkalis will not damage the fiber. With the exception of cool weak acids and mineral acids which will reduce the strength and eventually destroy the fiber completely. Organic solvents used in cleaning and bleaches, if handled properly, will not damage hemp.

Hemp fibers are used for cordage, rope, sacking, and heavy-duty covering fabrics. In some countries, hemp has been made into fine fabrics for use as wall coverings or draperies.

New Words and Expressions

jute 黄麻
burlap 粗麻布
hessian 打包麻布
grayish color 泛灰的颜色
fibrous 纤维状的
gummy 黏稠的
beige 米色
microorganism 微生物
bagging 打包布,麻袋布
carpet backing 地毯背用布
taping 贴边
linoleum 漆布
olefin 烯烃
labor-intensive 劳动力密集型的
ramie 苎麻
nettle 荨麻
family (植物的)科
perennial 多年生的
shrub 灌木
necessitate 需要
degum 脱胶
pectin 果胶
neutralize 中和
crystallinity 结晶度
non-fibrous 非纤维状的
hemp 大麻
deplete 耗尽
nutrient 营养
stripping 剥麻
drawing 牵伸
cordage 绳索
sacking 麻袋

Notes to the Text

1. since the dawn of the civilization:从文明开始时。
2. it did not attain economic importance until the latter part of the eighteenth century:直到18世纪后期才获得了经济上的重要地位。在英语中常用“it did not ...until”的句型表示一个肯定的含义。

3. it is used for bagging because it permits stacking without slipping or shifting position：它用做麻袋是因为盛装的物品不容易移动。指麻袋不容易变形。
4. It also is used as a base for the manufacture of linoleum："base"指"底布"。
5. It can be started from seeds：它可以用种子培育。
6. it can be grown from root cuttings：它可以用剪下的根培育。
7. reveal the fine fibers within：露出内部细的纤维。
8. Ramie reacts chemically in the same manner as other cellulosic fibers：句中"reacts chemically"可译为"在化学反应性能方面"。
9. low molecular accessibility：指外来物质不易进入其分子结构中。
10. bright and dark colors：浅色和深色。
11. heavy-duty：重型的。

Questions to the Text

1. Tell the history of the jute usage.
2. What are the chemical reactions of the jute fiber?
3. Why jute is suitable for bags?
4. How to grow ramie?
5. What are the properties of the ramie fiber?
6. Why ramie is highly resistant to microorganisms and insects?
7. What are the chemical behaviors of the hemp fiber?

Lesson Eight

Man-made Fibers

语言微课堂

Induction of man-made fibers

Actually the birth date of the"artificial silk" is said to date back some years before (1884) when an Englishman, Mr. Swan, produced small quantities of nitrocellulose which the researcher gad in mind to use for the development of incandescent bulbs. The first man-made fibers which were developed and produced used polymers of natural origin, more precisely of cellulose which is a raw material available in large quantities in the vegetable world.

Until the early years of the 20th-century, textiles were based solely on natural products such as cotton, wool, silk and linen. However, a chain of developments commencing in the 1890s led to a global textile industry based largely on fibers manufactured by industrial processes. These processes can be subdivided into those which use the cell walls of plants directly as their starting point, and thermoplastic fibers which start from non-renewable fossil reserves and hence use the plant matter indirectly.

Synthetic fibers, now economically manufactured from oil reserves valued at little more than the cost of extraction, have come to dominate the manufactured fiber market. According to the statistics, global natural fiber production in 2019 was over 27.6 million tons, the production in our country reached ca. 7. 3 million tons. The production and consumption of synthetic fibers is increasing at the expense of natural fibers and the improved perception of polyester, and the annual output of man-made fibers is close to 71 million tons.

Chemical fiber production trends in countries and regions are summarized in Table 8.1. The total production of the six main countries and regions has been close to 90% (2018) in the whole world, so it is possible to estimate global trends. China is the largest chemical fiber producer as well as consumer of filament/fibers utilized in textiles. The major portion of growth comes from Main land, China, 7. 7%-ish year-on-year growth represents the largest regional market for textile materials globally.

Table 8.1 The production capacity of synthetic fibers in main countries and regions (2018)

Countries/Reigon	Production capacity/kt	Year-on-year growth/%
Mainland, China	51, 964	7.7
Japan	633	−2.2

Continued

Countries/Reigon	Production capacity/kt	Year-on-year growth/%
Korea	1, 398	1.6
Taiwan, China	1, 661	-2.0
India	5, 467	-2.9
Europe	2, 457	4.6
Total	63, 579	6.1

* Exclude olefins, acetate fibers, Source: Chemical Fiber Production by Region from Japan Chemical Fibers Association

Chemical spinning technology

In the manufacture of rayon and other man-made fibers, the term "spinning" has come to be applied to the process of forcing liquid through tiny holes (spinnerets) which are able to solidify in a continuous flow to form the fiber. The spinning process is sometimes designated as "chemical or primary spinning" to distinguish it from the "textile or mechanical or secondary spinning".

The polymer processing from the solid to the fluid state can take place with two methods:

1. Melting process: This method can be applied on thermoplastic polymers which show stable performances at the processing temperatures (this method is used by 70% of the fibers).

In the case of melt spinning, the extruded polymer, owing to its fast cooling, is transformed directly into a filament while keeping substantially unchanged the form of the cross-section resulting from the filament geometry.

2. Solution process: The polymer is solved in variable concentrations according to the kind of polymer and of solvent, anyhow such as to produce a sufficiently viscous liquid (dope) (this method is used by 30% of the fibers).

In the case of solution spinning, the extruded filaments are subject to considerable structural changes brought about by the process for solvent extraction from the polymer mass.

Spinning via melting is definitely preferable as it entails a simple transformation of the physical state, however it can be applied only to polymer having a melting temperature (including PA, PET, PP); whereas spinning by solution is used in case that the polymers attain a thermal degradation at a temperature lower than a melting temperature (cellulose fibers, PAN). This last method is evidently more complicated than melt spinning, owing on one hand to the necessity of dissolving the polymer in a proper solvent, and on the other to the necessity of removing solvents and recovering the polymer after extrusion.

The flowchart which applies to the various kinds of spinning methods (Fig. 8. 1) is the following: the fluid polymer mass (melted or solution mass) is guided, through distribution lines, to the metering pumps (gear system), which guarantee a constant flow rate to the spinning positions, composed of a series of filters which purify and distribute the polymer; these are coupled with perforated plates of variable thickness and size, which are usually circular and made of special stainless steel (for melt spinning), but also of precious metals or of vitreous materials (for solution spinning).

The holes (capillaries), the number of which on the plate varies depending on the kind of fiber and can reach several thousands, can have circular or special cross-sections (shaped or hollow sections).

The filaments extruded from the spinnerets, after being converted back to their original state of solid polymer, are interrupted and taken up in suitable packages (bobbins, cans) or conveyed directly to subsequent processing phases.

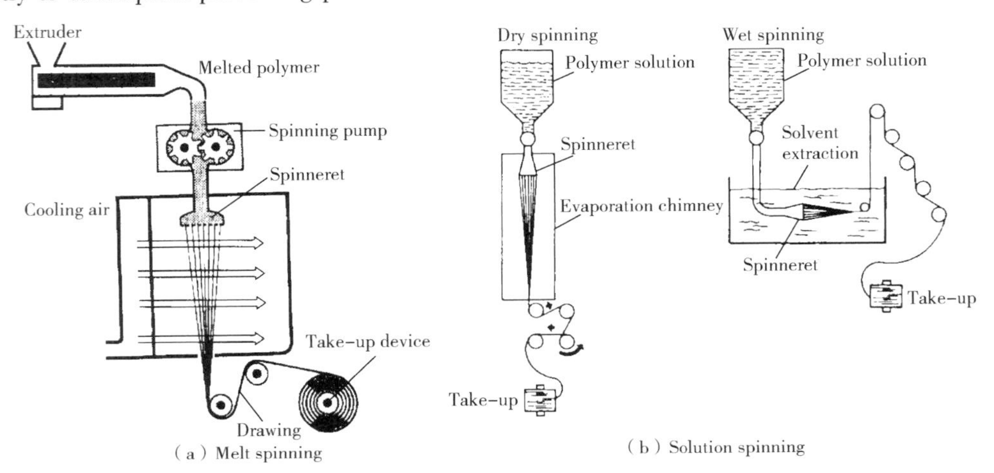

Fig.8.1 Spinning systems for man-made fibers

Solvent removal can take place in two ways:

Dry spinning

Solvent is removed through flows of warm gas suitably directed to the extruded filaments; gas temperature should be higher than the boiling temperature of the solvent, which will be extracted from the filaments, recovered and recycled.

专业微课堂：纺丝过程

Filament solidification proceeds according to the extent of solvent evaporation; it takes place faster on the external yarn layers (thus creating a crust or skin), and successively slows down while proceeding towards the interior.

As a consequence of the mass exchange, the original (round) cross-section of the filament undergoes a contraction, thus generating cross-sections which characterize the various kinds of fibers and spinning processes.

Wet spinning

This spinning method is based on the introduction of an extruded polymeric viscose into coagulation baths where the liquid, usually water, behaves as a solvent towards the polymer solvent and as a non-solvent towards the polymer mass.

Practically the solvent which is contained in the fiber in amorphous state (gel) is spread towards the liquid and at the same time the liquid of the bath is spread towards the interior of the fiber. The processing speeds are dependent on several parameters, as type and concentration of the polymeric solvent and of the liquid, which bring about structural variations in the fiber. In particular, the formation of an outer, gardened and more compact cortex (skin), similarly to what happens in dry spinning, slows down the coagulation mechanism of the inner filament portion (core), thus creating unevenness with a more or less porous structure (voids formation). The fiber cross-sections result more or less modified, from the original round form to a lobated form, with a wrinkled surface.

New Words and Expressions

nitrocellulose　硝化纤维，硝化纤维素
incandescent bulbs　白炽灯
subdivide　细分
little more than　差不多，仅仅是
perception　知觉，感知，洞察力
year-on-year growth　同比增长
olefins　聚烯烃
acetate fiber　醋酯纤维
spinnerets　喷丝头
dope　纺丝液
extrude　挤出
flowchart　流程表
distribution lines　分配板
dry spinning　干法纺丝
crust　硬皮
wet spinning　湿法纺丝
contraction　收缩
viscose　黏胶，黏液
amorphous　无定形的
lobated　叶状的，分裂状

Notes to the Text

1. Mr. Swan, produced small quantities of nitrocellulose which the researcher gad in mind to use for the development of incandescent bulbs：最初，Swan把硝化纤维通过小孔挤压而成细丝，再将这些细丝碳化，用于电灯泡。“gad”指“闲逛，游荡”，“gad in mind”指“脑海浮现”。
2. thermoplastic fibers which start from non-renewable fossil reserves and hence use the plant matter indirectly：热塑性纤维由不可再生的化石资源生产，因此可以说是间接由植物性的物质生产。“reserves”这里是名词，指化石资源。
3. Exclude olefins, acetate fibers：聚烯烃纤维、醋酯纤维除外。
4. solvent extraction：溶剂提取，这里译为“溶剂扩散”。
5. metering pumps (gear system)：计量泵。
6. perforated plates：有孔的板，这里指“纺丝组件内的喷丝板”。
7. precious metals or of vitreous materials：喷丝板的材料常由耐腐蚀的贵金属板或玻璃、瓷等制成。

Questions to the Text

1. Which countries contribute the man-made fiber production?
2. Describe the melt spinning technology.
3. Describe the solution spinning technology.

Lesson Nine

Regenerated Fibers *

Rayon/Viscose

Rayon is composed of cellulose. Like cotton it is a polymer. The difference in degree of polymerization (DP) between cotton and rayon accounts for some of the variance in physical properties between the two. Other causes for the difference between rayon and cotton include the difference in the degree of crystallinity, 60% for cotton and 40% for rayon, and differences in hydrogen bonding as evidenced by the infrared spectrum of native cellulose and regenerated cellulose.

The mechanisms of formation for natural celluloses are significantly different from man's ability to re-form cellulosic structures found in nature. Rayon is structurally different from cotton molecularly, morphologically, and in relation to chain length. Contaminants of various types from the pulp and other materials used in the process introduce significant differences to rayon that are not present in cotton. These differences result in significant differences in properties between cotton and rayon. As a consequence of the technological developments for regeneration of cellulose by man, a much wider range of fiber properties is obtainable from the man-made regenerated cellulose fibers than is possible from native celluloses such as cotton. A typical block diagram of the manufacturing process of rayon is shown in Fig.9.1.

To a significant extent, molecular orientation can be influenced by conditions used in the viscose rayon process. The current viscose process, however, in all of its variant modes to produce various rayon fibers, gives very little room for changing the crystallinity level of rayon fibers, which is approximately 40%, as compared with cotton, which is about 60%. Because of the relatively fixed degree of crystallinity in regenerated cellulose, the so-called amorphous areas are also rather constant.

Along the length or longitudinal direction, regular viscose rayon has a relatively uniform diameter and may also appear to have striations. These striations are the result of light reflection by the irregular surface contour arising from the cross-sectional shape of the fibers. A delustered fiber will have a grainy pitted appearance, while bright fibers appear mainly transparent. The cross section of regular rayons is highly irregular. Some may even have a bilobal appearance, a kidney bean shape.

The moisture regain for rayon is slightly higher than for natural cellulosic fibers. Because of this high absorbency, rayon fibers tend to dye and to finish more readily than cotton. The higher absorption is due in part to the lower DP; that is, there are more amorphous areas into which water, dyes, and/or finishes can be absorbed.

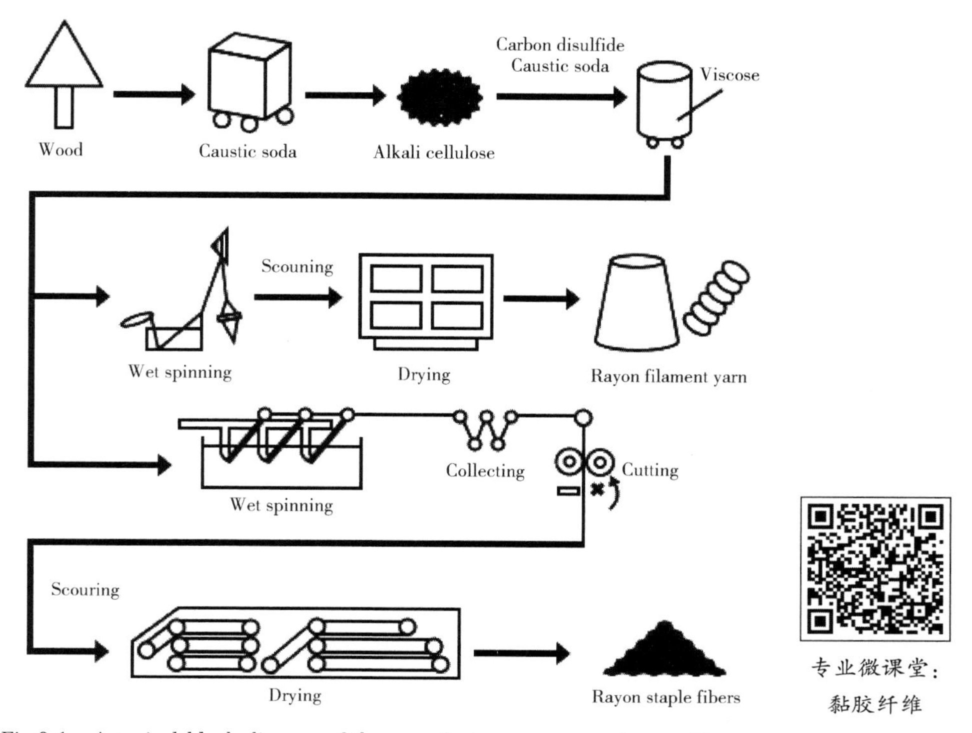

Fig.9.1 A typical block diagram of the manufacturing process of rayon fibers

Because it can be processed into either filament of staple form, rayon offers more variety in fabric and yarn construction than do the natural cellulose fibers. Through control of fiber size, yarn number, fabric construction techniques, dyes, and finishes, fabrics can be produced that are sheer to heavy, soft to firm, stiff to limp, in all colors including white. Simple, complex or textured yarns can be made.

Blends or combination fabrics available to the consumer included rayon and polyester; rayon and acrylic; rayon and cotton, rayon and flax. Rayon contributes absorbency and comfort when blended with other fibers; it also contributes styling, ease of dyeing, and softness. Rayon fabrics tend to stretch when wet and shrink upon drying. Rayon fabrics are stabilized against shrinkage by the use of chemical reactant finishes, or a combination of chemical reactant finishes and mechanical techniques.

Although laundry treatment and handling are dependent on yarn and fabric construction, finish and color application, and fiber content, rayon is not damaged by detergents and other laundry aids such as starches, fabric softeners, and water softeners. When it is essential for appearance, rayons can be bleached with hypochlorite or peroxide bleaches. However, it is essential that any bleach be properly diluted prior to the addition of fabrics.

Flame-resistant fabrics may require special care. Labels should be provided, and the directions given should be followed carefully in order to maintain the flame-resistant property.

Although rayon fabrics can be ironed at medium to high temperatures, finishes used on the fiber may require medium to medium-low temperatures. In fact, since rayon is cellulose, it can tolerate the same general care given to cotton fabrics of the same type and with the same finishes and dyestuffs.

Lyocell/Tencel

Lyocell is the first in a new generation of cellulosic fibers. The development of lyocell was driven by the desire for a cellulosic fiber which exhibited an improved cost/performance profile compared to viscose rayon. The other main driving force was the continuing demands for industrial processes to become more environmentally responsible and utilize renewable resources as the raw materials. The resultant lyocell fiber meets both demands. Lyocell was originally conceived as a textile fiber. The first commercial samples were produced in 1984 and fiber production has been increasing rapidly ever since. Fabrics made from lyocell can be engineered to produce a wide range of drapes, handles and unique aesthetic effects. It is very versatile and can be fabricated into a wide range of different fabric weights from women's lightweight blouse fabric through to men's suiting. Other end-uses, such as nonwoven fabrics and papers, are being developed.

Lyocell is a 100% cellulosic fiber derived from wood-pulp produced from sustainable managed forests. Typically, the DP of the pulp is in the range 400–1,000 units—Tencel® fibers have a DP of 500 to 550. The wood-pulp is pulled from the reels into a shredder which cuts the pulp into small pieces for mixing with the *N*-methyl morpholine oxide (abbreviated to NMMO or amine oxide) solvent. After the subsequent removal of excess water for making a homogenous solution with a minimum of undissolved pulp particles and air bubbles, the solution is then extruded (dry-spinning) into fibers and the solvent extracted as the fibers pass through a washing process in a water/amine oxide bath. The manufacturing process is designed to recover >99% of the solvent, helping minimize the effluent. The solvent itself is non-toxic and all the effluent produced is nonhazardous. The direct dissolution of the cellulose in an organic solvent without the formation of an intermediate compound differentiates the new generation of cellulosic fibers, including lyocell, from other cellulosic fibers such as viscose. This has led to the new generic name "lyocell" being accepted for labelling purposes.

High wet modulus rayons

There are two types of high wet modulus (HWM) rayons.

(1) Modal fibers, which were staple versions of the tougher tyre-yarn fibers, were introduced for use in industrial textiles, and for blending with the rapidly growing synthetics. The production of regular rayon, and of all the other types that have coagulation and regeneration followed by orientation, leads to fibers with low wet modulus. In contrast to this, the modal process, and others in which coagulation and stretch occur together and are followed by regeneration, leads to

high wet modulus. Modal fibers have the high tenacity, excellent luster as the silk and dry tactile feeling.

(2) Polynosic fibers is usually fabricated by cellulose with a high DP (minimal mercerizing of the alkali-cellulose). The high stretch oriented the cellulose molecules to a very high degree, giving the resulting fibers a high dry strength, an unusually high wet-to-dry strength ratio, a very high modulus and a characteristically high resistance to caustic soda. This latter point was regarded as important to allow the viscose fiber to be blended with cotton prior to the mercerizing (caustic) treatment, regular viscose being almost completely destroyed by this treatment. Polynosic fibers with even better wet stability and higher wet modulus were introdued to blend with and substitute for the better grades of cotton.

Others—Cuprammonium rayon and cellulose acetate fiber

Cellulose dissolves forming a cellulose-cuprammonium complex in cuprammonium solution. The cellulose-cuprammonium solution system can be traced back to that of the alcohol-alkali-copper system in 1898 by Fr. Billnheimer. The conventional cuprammonium process was clearly uncompetitive economically with the new viscose process and was inferior in quality to cellulose. The fundamental technological concept behind the wet-spinning of cuprammonium rayon is the complete separation of the following three steps, stretching, coagulation (solidification) and regeneration. This is essential to produce better quality fibers with high productivity.

Acetate fiber can be categorized into regenerated fiber, in fact, it is a semi-synthesized fiber. The starting material for acetate fibers was formerly cotton linters, the short fibers attached to the cotton seed, but wood pulp is now used. The purified fibers are steeped in glacial acetic acid, so that they become more reactive, and are then thoroughly mixed with an excess of glacial acetic acid and acetic anhydride. After that, the cellulose has been converted to cellulose diacetate or triacetate.

Fig.9.2 illustrates the schematic diagram of acetate fiber production based on dry-spinning technology fibers. The doping solution is extruded into the spinning cabinet. The emerging dope descends vertically for 2–5m against a countercurrent of hot air which enters the bottom of the tube. Finally, a guide roll at the bottom of the cabinet directs the fibers out to a collection device. The linear density of the filaments produced depends on the rate of mass flow through the spinneret holes, which is controlled by the metering pump, and the speed of take-up. The commonest filament sizes are around 3.7 denier. Tows, for cigarette filters or for cutting into staple fibers, are obtained by combining the output from a number of spinning positions.

The decline in textile uses of acetate was caused by the rapid expansion and price cutting of synthetic fibers, but unlike viscose the losses were more than compensated by the development of a massive new industrial market for the fiber: cigarette filter tips based on large crimped tows of diacetate. So far, despite many attempts by both synthetic and competitive cellulosic producers to displace it, diacetate tow reigns technically supreme in cigarette filters and worldwide production is

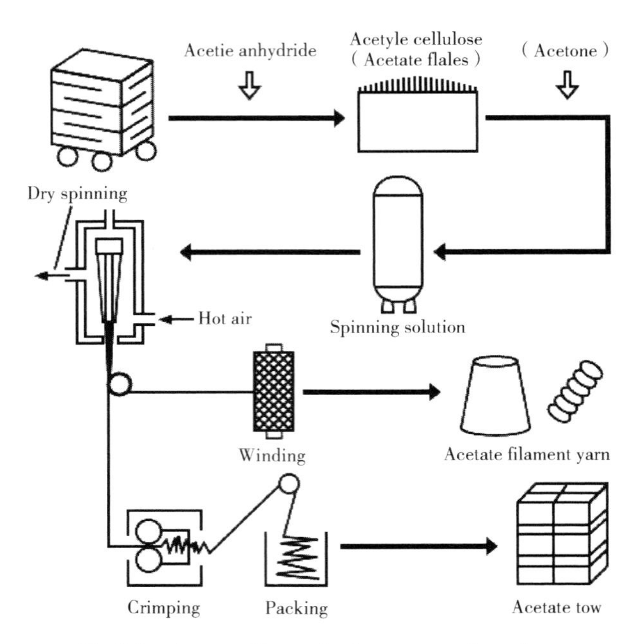

Fig.9.2 Acetate fiber spinning process flow chart

now estimated to exceed 800, 000 tonnes.

Acetate and triacetate fibers have a serrated cross-section, caused by solvent being lost from the interior after the surface has solidified by evaporation. Regular rayon, cuprammonium, and lyocell are round or circular in cross-section, and there are no irregularities in the contour. High-wet-modulus types come in several cross-sectional shapes-round, multi-lobal, and off round with a slight protrusion where the inside structure, or core, has penetrated the outside surface, or skin.

Currently, cellulosics, including cotton, may be losing market dominance to the cheaper synthetics, but nevertheless remain uniquely capable of providing levels of wearer comfort, absorbency, softness, biodegradability, and environmental compatibility that the synthetics have yet to emulate. These versatile regenerated cellulose fibers were tuned and meet the different market niches. Rayon fibers are extensively used in apparel and home furnishing fabrics; for industrial use such as reinforcing yarns, tires, and various types of reinforced rubber products, and for a wide range of nonwoven products, including personal care products, medical and surgical products, and a wide variety of wipes. High-wet-modulus rayon as a reinforcing medium for automotive tires is a premium performing material.

As regards regenerated fibers, it needs to be reminded that this group of fibers includes also fibers which have as raw materials natural polymers other than cellulose, like fibers derived from proteins.

A considerable historical significance was attained in Italy by protein fibers derived from casein, named Lanital, later on renamed into Merinova.

Protein fibers of animal origin (casein from milk) stopped to have commercial significance, whereas still to-day a certain interest is enjoyed, for example, in USA and China, by protein fibers of vegetable origin (maize, peanuts and soybeans).

New Words and Expressions

rayon 人造丝
degree of polymerization 聚合度
degree of crystallinity 结晶度
infrared spectrum 红外光谱
native cellulose 天然纤维素
regenerated cellulose 再生纤维素
mechanism 机理
morphologically 形态方面
contaminant 沾染物
pulp 浆粕
molecular orientation 分子取向
mode 模式，方式
amorphous area 非结晶区，无定形区
dyeability 染色性能
striation 条纹
delustered 消光的
bilobal 双叶形的
high-wet-modulus 高湿模量
off round 偏离圆形
surgical 外科用的
wipe 擦拭用品
limp 松软的
textured yarns 变形纱
acrylic 腈纶
reactant 反应物
starch 淀粉
softener 柔软剂
hypochlorite 次氯酸盐
peroxide bleach 过氧化物漂白剂
dyestuff 染料
profile 简介，特质
Lyocell/Tencel 莱赛尔/天丝
reel 卷
shredder 破碎机，碎纸机
amine oxide 氧化胺
effluent 废液
polynosic fiber 丽赛纤维，虎木棉，波里诺西克纤维，有的简称波诺纤维，属再生纤维素纤维的一种
tactile 触觉的
cuprammonium rayon 铜氨人造丝
cellulose acetate 醋酯纤维素
steep 泡，浸
glacial acetic acid 冰醋酸，冰乙酸
acetic anhydride 乙酸酐
serrated 锯齿状的
emulate 相仿，与……竞争
market niches 细分市场，nich，微环境
casein 乳酪
lanital （用酪素纤维制成的）人造羊毛

Notes to the Text

1. account for：是……的原因。
2. man's ability to re-form cellulosic structures found in nature：人改造天然纤维素结构的能力。
3. gives very little room：可选择的余地很小。
4. grainy pitted appearance：密布凹坑的外观。
5. there are more amorphous areas into which water, dyes, and/or finishes can be absorbed：句中"which"指"amorphous areas"；"finishes"指整理过程中加入的化学成分。

6. in all developed parts of the world：所有发达国家。
7. High-wet-modulus rayons have less potential for this than other rayon fibers：句中"this"指"stretch when wet and shrink upon drying"。
8. mechanical technique：指物理整理。
9. laundry treatment：指洗涤方式。
10. prior to the addition of fabrics：在加入织物之前。
11. This has led to the new generic name "lyocell" being accepted for labelling purposes："Lyocell"本义由"lyo"和"cell"组成，溶解分散的细胞。

Questions to the Text

1. Why rayon and cotton have different properties?
2. Why higher moisture regain of rayon is considered to be a good property?
3. Describe the usages of the rayon fiber.
4. What are the differences between rayon and lyocell?

Lesson Ten

Synthetic Fibers

语言微课堂

Polyester fibers

Polyester is the most common of all the man-made fibers and second most popular fiber after cotton.

Polyester fibers were first identified by Carothers and his team at DuPont during the early years of their fundamental research. However, when polyamide fibers (nylon) appeared to show promise, they were selected for development, and research on polyester fibers was temporarily set aside.

While Carothers and his team decided their emphasis to the polyamides, chemists in Great Britain began experimenting with long chain linear polyester polymers. In 1941 Calico Printers Association introduced a successful polyester fiber. Its development was delayed by World War Ⅱ, and public announcements of the discovery were withheld until 1946. Imperial Chemical Industries (ICI) purchased the rights to manufacture the fiber for all countries except the United States, where DuPont obtained the manufacturing privilege. Today, polyester fibers are manufactured most abundantly among synthetic fibers. The amount of production is continuing to increase year by year. Polyester fibers have such features as extreme strength, which varies from a low of 2.5 grams per denier to a high of 9.5 grams per denier, resistance to crease, less moisture absorption, thermoplasticity, and resistance to acid and alkali.

The staple fibers are friendly with other types of fibers. They are manufactured into fabrics, by blending with cotton, wool and linen fibers, making the use of each characteristic. The staple fibers are largely used for wadding as they are. By modifying their characteristics, anti-static fibers and flame-retardant fibers are also manufactured.

End-uses

(1) For clothing. Women's and children's wear, men's wear, pupils' uniforms, lining, rain coat, dress shirts, blouses, working wear, neckties, socks, sweaters, knitted underwear, sports wear, etc.

思政微课堂：
合成纤维

(2) For home furnishings and bedding. Curtain, table cloth, blankets, wadding for bedding, sheets, etc.

(3) For other uses. Umbrellas, sewing threads, synthetic leather, artificial leather, etc.

Features

(1) It is a fine, regular, and translucent fiber (Fig.10.1).

(2) One of the strongest fibers and no deterioration is remarked even if the fibers are wetted.

They exhibit the same characteristic in abrasion. In addition, they are considerably resistant to heat.

(3) Since superior in crease recovery, difficult to de-form the shape.

(4) Little deterioration in strength occurs even if polyester fibers are exposed under sun light for a long period of time.

(5) Stability in dimensional change, fast drying, and little wrinkling after laundering due to a little moisture absorption needs little ironing.

(6) Pleats and creases cannot be removed after laundering due to thermo-plasticity.

(7) Resistant to chemicals and non-attackable by insects and molds.

Care instructions

The redeposition may occur if washed with heavily dirty cloth, or soaked in washing solution for a long period of time. The redeposition is a phenomenon that clean fibers absorb the dirt resulting in dark hue. When washing clothes of white or light color, avoid washing with other dirty clothes and try to wash for a short period of time.

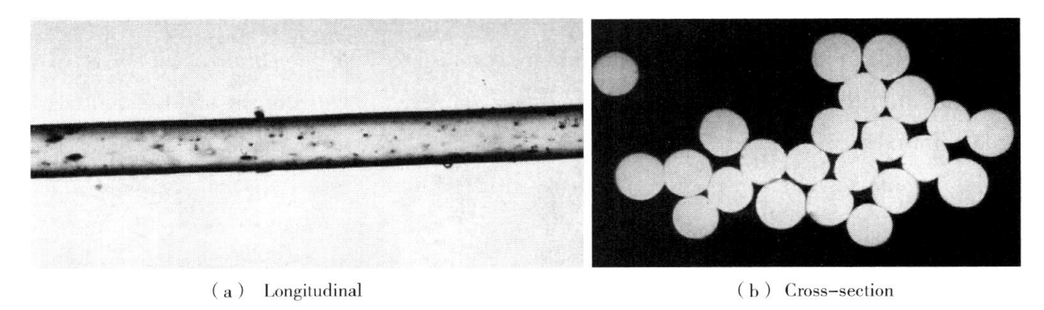

(a) Longitudinal (b) Cross-section

Fig.10.1 Longitudinal and cross-section view of polyester fiber

专业微课堂：纤维的分类

Acrylic/Modacrylic fibers

Acrylic fibers have characteristics most similar to those of wool fibers. They are soft and light fibers having light and warm tactile feeling to the human skin. Acrylic fibers are widely used, making the most of their characteristics, for knitted products such as sweater and jersey; bedding textiles such as blankets and pile sheets; and carpets for home-uses. In addition, rag animals and western wigs are the products making use of acrylic fibers.

Acrylic fibers are manufactured in a wide variety of characteristics such as those lustered, delustered, with soft tactile feeling, and with rigid tactile feeling. There are many other types of acrylic modified fibers such as animal-like fibers with oval cross-section, thermal resistant fibers, anti-pilling fibers, antibacterial and deodorant fibers, and anti-static fibers.

Acrylic fibers are manufactured in staple fibers, almost with a wet spinning process. However, some fibers are manufactured with a dry spinning process. Such dry spun fibers have different fabric hand from that of wet spun fibers and are developing in a variety of new end-uses. Light/soft feeling, a characteristic among features of acrylic fibers, is referred to as bulkiness. Bulky spun yarns, difficult to spin other fibers, are available. A variety of yarns blended with other types of fibers such as cotton and wool fibers are manufactured to make use of characteristics of each type of fibers.

Filament yarns of acrylic fibers are also available. Since the filament yarns have luster and tactile feeling, they are used for silk-like knitted products. In addition, they are used for accessories for national dresses and the field of silk products thanks to the non-yellowing. For information, acrylic fibers include acrylic fibers themselves and modacrylic fibers. Those fibers which contain (not less than 85%) acrylonitrile molecule as main component are "acrylic" fibers. On the other hand, those fibers which contain 35 to 85% acrylonitrile molecule are "modacrylic" fibers. Since modacrylic fibers are copolymerized with vinyl chloride or vinylidene chloride, they are flame retardant. They are used for curtains and wigs.

End-uses

(1) For clothing. Sweaters, women's and children's wear, sports wear, socks, knitted underwear, pajamas, gloves, etc.

(2) For home furnishings and bedding. Carpets, a variety of rugs, upholstery, cushions, blankets, pile sheets, etc.

(3) For industrial uses. Felts for paper making, filter cloth, alternative asbestos, tents, sheet, etc.

(4) For other uses. Rag doll, toys, auxiliary tapes for bags, braids, cloth for bags, wigs, etc.

Features

(1) Lighter than wool fibers, and fabric hand with bulkiness.

(2) Superior in warm retention, light and warm.

(3) Superior in elastic recovery, and resistant to crease.

(4) Excellent in color development and can be dyed in desired color.

(5) Little affected by sunlight.

(6) Resistant to chemicals and cannot be attacked by molds and insects.

Care instructions

In the laundering, care should be taken not to de-form the shape. In particular, wash bulky knitted products with hands, fully squeeze with bath towel, and screen dry in a well ventilated place.

Since products of acrylic fibers are particularly sensitive to heat, avoid to use a tumble dryer (hot air dryer). If the product is directly ironed at a high temperature, the product may damage its fabric hand or may shrink with heating. The products should be ironed with a floating steam iron to restore the shape.

Nylon fibers

Nylon fibers are one of the fibers having the longest history among synthetic fibers. They are used for a variety of uses from clothing to home furnishings and industrial uses.

In nylon fibers, there are many types depending on the raw materials. The fibers most manufactured in China"Nylon 6" fibers. In addition, "Nylon 66" fibers, rather resistant to heat, are also manufactured. Comparing other synthetic fibers, great advantages of nylon fibers are resistant to abrasion and flexing, and of supple tactile feeling. Accordingly, thin, light and flexible woven or knitted fabric can be manufactured. Furthermore, an advantage of easily dyeability makes possibility recent year to develop many fabrics for clothing.

Nylon fibers can be produced in a variety of cross-section and fineness. Composite fibers, having unique appearance and tactile feeling, can also be produced by combining with other types of fibers. Heat storage or warmth retention fibers consisting of extremely fine filament yarns in which carbonaceous material converting the light to heat is inserted. Anti-static nylon fibers inhibit electro-static charge. Thus a wide variety of nylon fibers are available giving much more performances and fabric hand.

Almost all nylon fibers are filament yarns ranging from very fine to coarse denier to be used for clothing such as rain wear, sport wear, lingerie, pantyhose; home furnishing as carpets; and industrial uses such as fishing nets, substrate for synthetic leather, ropes, and tire cords.

Staple fibers are used, in the yarns blended with wool and acrylic fibers, for clothing; home furnishing such as carpets, upholstery; and other fields of sundries.

Features

(1) One of the very strong fibers. Extremely resistant to abrasion and flexing.

(2) The specific gravity is 1.14. Very light, i. e., 80% of that of silk fibers, and 70% of that of cotton fibers.

(3) Since nylon fibers absorb little water even though they are wetted, they dry fast and simple in laundering.

(4) Excellent in elasticity and resistant to wrinkle.

(5) If properly thermo-set, nylon textiles little shrink/extend or little de-form their shape due to thermoplasticity.

(6) Resistant to chemicals and oil. Non-attackable by sea water.

(7) Non-attackable by molds and insects.

Care instructions

A white cloth of nylon fibers tends to yellow gradually, not so severe as silk fibers, by exposing ultraviolet ray (sunlight). To prevent this phenomenon, a variety of countermeasures have been made. A white cloth should be dried in the shade. Nylon fabrics should be ironed at the low temperature.

Vinylon fibers

The fiber invented in Japan and grown world-wide. A vinylon fiber is said to be the most similar to cotton fibers since it is most moisture absorbent among synthetic fibers. Vinylon fiber is widely used for clothing, such as substrate for lace, working wear, etc. , as well as for industrial, agricultural, fishing uses, because it is light in weight, durable and resistant to weathering. Like the cotton, the tent/canvas made by vinylon can breathe, which provides good resistant to the water when wet and air permeability when dry.

Although vinylon fibers is produced mainly in staple fiber, it is also produced in filament yarn. Fabric hand of some filament yarn is similar to that of silk yarn.

Features

(1) The specific gravity is 1.26–1.30. Vinylon fiber is lighter in weight than rayon, acetate, wool and cotton fibers.

(2) Especially, excellent in resistance to abrasion.

(3) Most similar to the tactile feeling of cotton fiber and most moisture absorptive among synthetic fibers.

(4) Resistant to acid and alkali. Difficult to rot. Non-attackable by molds and insects.

Care instructions

Vinylon products may become rather hard or yellow, if ironed in wet condition. Please iron after complete drying.

Polypropylene fibers

Polypropylene fibers are manufactured by polymerization of propylene produced by purifying the petroleum. Polypropylene fiber is lightest in weight among synthetic fibers. It can float in water. Polypropylene fiber is used for home furnishings such as carpets and rugs; and industrial uses such as ropes and filter cloth, reflecting the characteristics of strong tensile strength, resistance to acid and alkali and soil resistance. In addition, it is also used for swimsuits, socks and underwear with superior in warmth retention and fast drying properties thanks to little moisture and water absorption. On the contrary, it is inferior in heat resistance.

Self-adhesive composite fibers compounded of polypropylene fiber and other polymers of low melting point are also manufactured for a wide variety of hygienic products, medical supplies including medical masks.

Features

(1) The specific gravity is 0.91, lightest in weight among all fibers. One of the very strong fibers.

(2) Little moisture and water absorption. Quickly dryable even it is wet. Superior in warmth retention thanks to the low heat transfer rate.

(3) Resistant to acid and alkaline chemicals.

(4) Thermoplastic.

Care instructions

Wash with water avoiding the dry cleaning. Avoid tumble drying (hot air dryer), iron at low temperature and use an iron cloth, because polypropylene fiber is heat sensitive and tends to be de-formed by heating.

Polyurethane fibers

Polyurethane fibers are referred to as "Spandex" as a generic name. The fiber itself is highly stretch and elastic as rubber, strong than rubber, and resistant to aging. The fiber is dyeable freely and fine yarn can be produced.

Polyurethane fiber is manufactured from glycol and diisocyanate as raw material with a special spinning method. No product is manufactured of 100% polyurethane fiber. The polyurethane fiber is used by blending other fibers. A fabric, referred to "two-way tricot" in which a polyurethane fiber is knitted with a nylon or polyester filament yarn, is highly elastic and it is used for swimsuits, leotards, underwear, bandages, paper diapers and other similar stretchable clothing.

A yarn, in which another yarn such as nylon yarn is wound around polyurethane yarn is referred to as a "covered yarn". A covered yarn is used for pantyhose and foundation garments. In addition, a yarn in which a polyurethane yarn is inserted as a core in the spinning process is referred to as a "core spun yarn". A core spun yarn is used for tights and training wear.

Features

(1) Extends 5-7 times when stretched as rubber.

(2) More durable than rubber and 2-4 times higher tensile strength than rubber. More resistant to aging than rubber and finer yarn is available.

(3) Since considerably resistant to high temperature, a heat setting can be achieved with other fibers simultaneously.

(4) Although non-attackable with alkali, yellowing with the action of chlorine and deterioration is accelerated with chlorine.

Care instructions

Avoid chlorinated bleaching agents. Do not iron by stretching a polyurethane fabric.

Polyvinyl chloride fibers

Polyvinyl chloride fibers are invented in Germany in 1931 and earliest fibers among synthetic fibers. Vinyl chloride is polymerized into polyvinyl chloride and is solution-spun into polyvinyl chloride fibers. Durable and resistant in weathering, non-attackable by acid and alkali and superior in warmth retention property. By rubbing, polyvinyl chloride fibers are charged with negative static electricity. Accordingly, it is said that the wearing of underwear of this fiber will be effective against rheumatism, and underwear of polyvinyl chloride fiber is available as a healthy underwear. However, this fiber is less resistant to heating. Filament fiber yarn starts to shrink from ca. 60% and stable fibers start to shrink from ca. 90 to 100%. Thus polyvinyl chloride fibers are less used for clothing and wigs which needs ironing. For industrial uses, polyvinyl chloride fibers are used for coverings of an electric wire due to excellent electrical insulation.

Features

(1) Extremely resistant to chemical agents and sunlight.

(2) High thermal and electrical insulation.

(3) Chargeable with negative static electricity.

(4) Does not absorb water completely.

(5) Highly bulky and superior in warmth retention.

(6) Flame retardant.

(7) Most care should be taken for heat, because of low resistance to heating.

Care instructions

Do not pour hot water, or do not dry near a stove. Do not iron clothing of general polyvinyl chloride fibers.

Polyethylene fibers

Fibers manufactured from polyethylene produced by polymerizing ethylene. Although polyethylene fibers are very strong but poor in heat resistance. Polyethylene fibers are not used for clothing but used for industrial uses, such as insect-proofing nets, cords and strings, fishing nets, ropes, filter cloth, etc.

High-tenacity polyethylene fibers produced with a special manufacturing procedure are three times strong than (high-tenacity) nylon fibers and are used for an anchoring rope.

Features

(1) Strong, resistant to chemical agents, little water and moisture absorbent.

(2) Light (0.94–0.96g/cm^3 in specific gravity) fibers next to polypropylene fibers.

(3) Poor in thermal resistance. Follow the same care instructions with polyvinyl chloride fibers.

Words and Expressions

artificial leather 人造皮革
jersey 针织套头衫
deteriorate 恶化
flame-retardant 阻燃
redeposition 再沉积
pile 堆，绒头
rag 碎布
wig 假发
vinyl chloride or vinylidene chloride 氯乙烯/偏氯乙烯
rug 厚毯
upholstery 家具装饰品
asbestos 石棉
flex 屈伸
supple 灵活的，柔顺的
lingerie 女内衣
pantyhose 连裤袜
sundries 杂项
glycol 乙二醇
diisocyanate 二异氰酯
two-way tricot 双向经编织物
leotards 紧身连衣裤
diaper 尿片，纸尿裤
wadding 衬垫
stove 炉子

Notes to the Text

1. appeared to show promise：似乎很有前景。
2. set aside：搁置。
3. Calico Printers Association：英国一印染公司。
4. In particular, wash bulky knitted products with hands, fully squeeze with bath towel, and screen dry in a well ventilated place. 特别地，洗蓬松的针织产品时，需要手洗、用浴巾辅助拧干、保持平整状态在通风良好的地方晾干。"screen dry"是指"平整干燥"。
5. Heat storage or warmth retention fibers：蓄热纤维。

Questions to the Text

1. Among the synthetic fibers, which has the best elasticity? Which is the lightest? Which belongs to fire-retardant fiber?
2. Please talk about the main features of the specific synthetic fiber in text? (choose 3 or more)

Lesson Eleven

Functional Fibers (1) *

语言微课堂

专业微课堂：电纺抗菌纳米纤维

Antimicrobial fibers

As consumers are becoming increasingly aware of the implications on personal hygiene and the health risks associated with some microorganisms, the demand for antimicrobial textiles has presented a big increase over the last few years. Antimicrobial fibers and their products have a variety of applications. For example, socks that claim to be antimicrobial can reduce the reproduction of fungi fed by human sweat and solve the unpleasing feet odor problem. Mildew-resistant shower curtains are another example of antimicrobial textiles as they have resistance against the growth of a type of fungus. Besides antifungals, textiles can also be combined with antibacterial agents and have a wide market in clinics, such as sterilized bandages.

There are different chemical and physical approaches that have been developed or that are under development to impart antimicrobial properties to the textile fibers. One approach to develop textiles with an antimicrobial ability is by the incorporation of the antimicrobial agents into the polymeric matrix of the textile fibers. The direct addition of the biocide agent into the polymers has received considerable attention, especially when using thermoplastic matrices. The main advantage of this method is that it may be easily implemented in the standard and large-scale processing units already designed to prepare particulate-filled polymer composites. The main disadvantage of this method is the lower antimicrobial power due to the restricted diffusion of the antimicrobial agent molecules through the polymeric matrix.

Another approach, which can be used for synthetic and natural fibers or any textile fabric, is the application, in the finishing stage, of antimicrobial agents on the material surface. Over the past decade, several surface grafting techniques have been studied; however, the method used strongly depends on if the textile fiber is natural or synthetic and also on its physico-chemical features. Different techniques have been used to achieve textiles surface grafting, such as: ①chemical grafting; ②plasma-induced grafting using either radiofrequency or microwave plasma; ③radiation-induced grafting, which uses high-energy radiation; ④light-induced grafting using a source of ultraviolet radiation.

Phase change fibers

Phase change fibers can regulate temperature by allowing one or more components of the fiber to melt while one or more other components help maintain the structural integrity of the fiber. Thus, energy is used to melt a portion of the fiber, rather than being used to raise the temperature. The temperature regulation range depends on the melting temperature and on the enthalpy of melting of the component used for this purpose. The working principle of phase change fibers can be explained as follows. Phase change fibers will react with the changes of environmental temperature and the temperatures in different areas of human body. Phase change fibers absorb body heat when the body temperature increases and phase change component melts. However, melted or semi-melted phase change component gives off heat when body temperature decreases. In this way, phase change fibers could keep the body in a comfortable state and are used for the thermal management of garments. The process of phase change is a dynamic; hence, the phase change component inside the fibrous structures is constantly changing from solid to liquid and vice versa depending upon the physical activity of person and external environmental temperature.

Some approaches to manufacturing phase change fibers are (a) incorporation of phase change materials (PCMs) into the fiber core, (b) incorporation of PCMs into fibers as a repeating unit in a block copolymer, (c) coating of PCMs on the surfaces of fibers, and (d) use of PCMs alone as an independent polymer, as shown in Fig.11.1. Applications of phase change fibers include shoes such as ski boots, mountaineering boots, and race car drivers' boots; sportswear; bedding accessories such as quilts, pillows, and mattress covers; and space suits. Polyethylene glycol (PEG) is a commonly used as PCM in fibers. Different phase change fibers such as cellulose acetate and polyvinylidene fluoride (PVDF) that contain PEG have been developed.

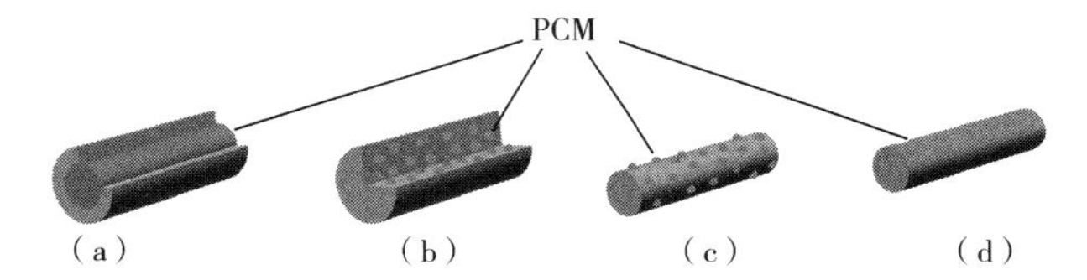

Fig.11.1 The types of phase change fibers

Biomedical fibers

Fibers are being used for many biomedical applications, including as scaffolds for tissue engineering, in drug delivery, in wound dressings, as biosensors, and as potential surgery tools. Scaffold in tissue engineering acts as an artificial extracellular matrix that provides a potential growth site for cells, and it helps maintain tissue structure. These scaffolds will be used in the body, so the materials need to be biocompatible. Some examples of polymers used for scaffolds are polylactic acid (PLA), poly (lactic-coglycolic acid), poly-L-lactic acid (PLLA), and polycaprolactone (PCL). Electrospinning is commonly used for making fibers for scaffolds. Scaffolds with large specific surface area are desirable, as they provide more potential sites for functionalization and cell growth. The purposes of wound dressing are to prevent the buildup of exudate and to keep the wound clean, increasing recovery rate and preventing infection. Similar to the case for scaffolds' fibers, electrospinning is currently the most common way to produce fibers for wound dressing. The porous structure of electrospun fiber mats makes the wound dressing semipermeable. Polymers studied for wound dressing include polyurethane, gelatin, poly(vinyl alcohol), collagen/chitosa, and silk. A wound covered with electrospun nanofibrous polyurethane wound dressing had less inflammation and higher rates of wound healing than did a wound covered with a commercial polyurethane film.

Silver nanoparticles (nAg) were incorporated into fibers to improve the antibacterial properties of the electrospun scaffold and wound dressing. PLA/nAg electrospun fibers had an antibacterial efficiency of as high as 98% against Staphylococcus aureus and of 94% against Escherichia coli. Other polymers spun with nAg also have enhanced antibacterial properties compared to neat fibers without nAg.

Drug delivery through fibers has also been explored. The most common processing method for these fibers is also through electrospinning. There are three main ways to incorporate drug delivery capabilities into fibers: coating the fiber with the desired drug; blending the drug in the polymer solution prior to fiber spinning; and bicomponent spinning, in which the drug is usually in the core and a polymer is used as the sheath. Many different types of drugs in different polymer matrices have been incorporated. The dissolution/delivery of the drugs is dependent on multiple factors, including the surface area of these fibers, the polymer matrix, and the geometry of the fiber.

New Words and Expressions

antimicrobial 抗菌的
microorganism 微生物
biocide agent 杀菌剂
physico-chemical 物理化学的
radiofrequency 射频
temperature regulation 体温调节
block copolymer 嵌段聚合物
space suits 太空服
scaffold 支架
electrospinning 静电纺丝
inflammation 发炎
antibacterial 抗菌的
Staphylococcus aureus 金黄色葡萄球菌
Escherichia coli 大肠杆菌

Notes to the Text

1. The temperature regulation range depends on the melting temperature and on the enthalpy of melting of the component used for this purpose: "enthalpy of melting"是指"熔融焓"。
2. In simulated tests: 在模拟实验中。
3. Some examples of polymers used for scaffolds are polylactic acid (PLA), poly(lactic-coglycolic acid), poly-L-lactic acid (PLLA), and polycaprolactone (PCL): 用于支架的聚合物有聚乳酸、聚乳酸-羟基乙酸共聚物、聚左乳酸和聚己内酯。
4. Polymers studied for wound dressing include polyurethane, gelatin, poly(vinyl alcohol), collagen/chitosa, and silk: 用于伤口绷带包括聚氨酯、明胶、聚乙烯醇、胶原蛋白/壳聚糖和丝织物。

Questions to the Text

1. What approaches are usually used to impart an antimicrobial ability to textiles?
2. What is the mechanism of phase change fibers in the temperature regulation?
3. What are the approaches to manufacturing phase change fibers?
4. What materials can be used for manufacturing biomedical fibers?
5. List the types of functional fibers you know and their potential applications.

Lesson Twelve

Functional Fibers (2) *

语言微课堂

Energy harvesting fibers

Fibers can be used to harvest energy from a multitude of sources by utilizing piezoelectric properties, sunlight, and mechanical movement. Piezoelectric fibers have higher flexibility than some piezoelectric bulk materials, and thus these fiber composites can be used in a broader range of environments. A PVDF/sodium niobite ($NaNbO_3$) nonwoven fabric generated 3.4V voltage and 4.4μA current in a cyclic compression test that simulated conditions of a human body walking. This fabric also had good durability and showed good properties after a million compression cycles and is thus a good candidate material for a wearable power generator. Poly(tetrafluoroethylene) fibers were used in triboelectric nanogenerators, which generated power from mechanical motion. This hybrid power textile can continuously power an electronic watch, can directly charge a cell phone, and can drive water-splitting reactions.

A polymer fiber-based fabric that can simultaneously harvest energy from both sunlight and mechanical movement was developed. In that study, ZnO/nanowire arrays grown on manganese plated poly(butylene terephthalate) (PBT) fibers with dye sensitization were used as photoanodes, and copper-coated PBT fibers were used counterelectrodes in solar cells, which generated power from sunlight.

Shape deformable fibers

专业微课堂：
形状记忆纤维

Shape deformable fibers can reversible change its position or shape in response to external stimuli, such as magnetic field, electricity, irradiation, heat, and atmosphere. Design and construction of controllable shape deformable fiber device is the key to realize the development of deformable clothing. Researchers have devoted a lot to the study of shape deformable fibers, and have made great progress. According to external stimuli, shape deformable fibers can be divided into several categories, such as electrically controlled, solvent-responsive, and light-induced deformable fibers. As for electrically controlled deformable fibers, there are usually two ways to obtain electro-deformed fibers: one is dielectric elastomer-based electrically controlled deformable fibers, which drive the elastic deformation of the dielectric polymer by high voltage; the other is to drive the fiber deformation by Joule heat generated by electric energy. For the solvent-responsive deformable fiber, solvents are usually adsorbed and desorbed on the surface of the fiber to drive its deformation which is the direct way to change to shape or length of a fiber. Light inducing is a non-contact method to achieve material shape change, which has long-distance controllability. The above deformable fibers not only show simple rotation and bending, but also can be knitted into ordinary fabrics to drive fabric deformation.

Electromagnetic shielding fibers

Currently, electrically conductive fibers are very often used for electromagnetic (EM) shielding applications because natural or synthetic fibers, for example, cotton, polyester, polyamide, polyacrylic and cellulose acetate fibers, exhibit very poor electric conductivity. There are many metal-coated textile fibers, metal fibers, carbon based fibers, etc., available on the market. The surface coating is often time consuming, laborious and costly, whereas the spinning process of conductive filler-reinforced polymer composites is more efficient. The conductive fillers in the form of particulates, fibers and filaments are increasingly being used for shielding of EM radiation. Conductive filler particles, such as silver, copper, gold, etc., and carbon particles, such as carbon nanotubes, are used for EM shielding applications in many studies, giving excellent results. Conductive polymers, on the other hand, allow an excellent control of the electrical stimulus, possess very good electrical and optical properties, have a high conductivity/weight ratio and can be made biocompatible, biodegradable and porous. The conductive polymers, such as polypyrrole (PPy), polyaniline, polyacetylene, etc., have been described for their EM shielding applications in many scientific studies.

Chromatic fibers

In recent years, clothes with constant color can no longer meet people's demand for fashion and novelty. Therefore, flexible intelligent chromatic fibers have become a hot topic and have been widely studied. According to the different physical or chemical mechanisms, chromatic fibers can be divided into electrochromic, thermochromic and structurally colored fibers.

The sandwich-like structure is considered to be a typical structure of electrochromic devices, which consists of two transparent conductive electrodes with an internal electrochromic active layer. Therefore, electrochromic fibers are fabricated by spirally rolling the narrow electrochromic films, paralleling the coil electrode on a fiber substrate, and wrapping fiber electrode on a fiber device. These electrochromic fibers could change color quickly upon voltage change.

Thermochromic fibers are made of compounds or mixtures that change their visible absorption spectra when heated or cooled. The thermochromic fibers have the characteristics of discoloration at a specific temperature, showing a new color, and restoring to the original color when the temperature is restored to the initial temperature. There are two main ways to prepare thermochromic fibers: composite fibers and surface coatings. The mechanical properties of composite thermochromic fibers are usually poor, which makes it difficult to meet the requirements of weaving.

Structurally colored fibers refer to a kind of fibers with color on the surface or in the interior due to periodic structure. Different from the traditional fiber color, the color of structurally colored fibers is mainly produced by the interaction of micro-nanostructure and light on its surface or inside. The color of the structurally colored fibers can be changed by adjusting its inner or surface periodic structure. In addition, there is almost no energy consumption in the process of color changing of structurally colored fibers. Therefore, the preparation and application of structurally colored fibers have attracted extensive interest of researchers.

New Words and Expressions

energy harvesting 能源采集
piezoelectric 压电
sodium niobite 铌酸钠
poly(tetrafluoroethylene) 聚四氟乙烯
water-splitting 水解
manganese 锰
photoanode 光阳极
counterelectrodes 对电极
solar cell 太阳能电池
electromagnetic shielding 电磁屏蔽
polypyrrole (PPy) 聚吡咯
polyaniline 聚苯胺
polyacetylene 聚乙炔
chromatic fiber 变色纤维

Notes to the Text

1. Fibers can be used to harvest energy from a multitude of sources by utilizing piezoelectric properties, sunlight, and mechanical movement: “a multitude of”是指“众多”。
2. ZnO/nanowire arrays grown on manganese plated poly(butylene terephthalate) (PBT) fibers with dye sensitization were used as photoanodes: “manganese plated poly(butylene terephthalate) (PBT) fibers”是指“镀锰的聚对苯二甲酸丁二醇酯纤维”。
3. ...the key to realize the development of deformable clothing: “the key to realize”是指“实现…的关键”。
4. Structurally colored fibers refer to a kind of fibers with color on the surface or in the interior due to periodic structure: “structurally colored fibers”是指“通过结构设计的变色纤维”; “periodic structure”是指“周期性结构”。

Questions to the Text

1. List the types of smart fibers you know and their potential applications.
2. What are the sorts of the energy harvesting fibers?
3. How many categories can shape deformable fibers be divided into?
4. Describe the mechanisms of chromatic fibers.

Lesson Thirteen

Fiber Identification

语言微课堂

The identification of fibers is critical to a variety of industries, including textiles, forensic science, fashion, and design. Changes in textile technology create a constant need to improve identification methodology. Qualitative identification of textile fibers can be difficult, and it may require several tests. Simple tests that can be used in identification are described here along with brief comments about their use and importance. In addition to their use in fiber identification, some tests yield insight into problems of processing and care of textile products.

Much of the information presented here is useful in daily life. A consumer with a certain amount of training in textiles will find the ability to tentatively identify fibers extremely valuable. People may wish to verify label information and may need to know what fibers or fiber groups are in a fabric in order to care for it properly.

The major test described include the burning test, which helps place a fiber into categories but seldom identifies a specific fiber. Microscopic evaluation, which is a little more specific and in some cases may be accurate enough to identify individual fibers; and chemical solubility, which may be accurate enough to categorize fibers into generic groups. Instruments used in more precise and accurate identification are mentioned. Readers who wish to be able not only to state what fiber type is involved but to identify a specific fiber would usually need instrumental evaluations as well as those described herein.

The burning test

The burning test is a good preliminary test. It provides valuable data regarding appropriate care and will help place a fiber into a specific category. It is not, however, a test that can be used alone to provide exact identification of specific fibers. In the case of yarns composed of two or more fibers, the test will usually give the reaction of the fiber that burns most easily; if a fiber is heat sensitive, it will tend to melt or withdraw from the flame, leaving the flammable fiber to burn. Remember that the burning test is a preliminary test that indicates general grouping or categories only. The test is relatively simple but must be used with care to avoid injury and guard against fire. The procedure to follow in doing the burning test is as follows:

(1) Select one or two yarns from the warp of the fabric or unravel a length of yarn from a knitted structure.

(2) Untwist the yarn so that the fibers are in a loose mass.

(3) Hold the loosely twisted yarn in forceps; move them toward the flame from the side (That is, approach the flame from its own level, not by bringing the sample down into the flame).

(4) Observe the reaction as the yarn approaches the flame.

(5) Move the yarn into the flame, and then pull it out of the flame and observe the reaction. Does the yarn start to burn as it nears the flame? Does it start to melt? Does it shrink away from the flame? Does it burn quickly or slowly? Does it have a sputtering flame, a steady flame, no flame at all? When removed, does it continue to burn? Is it bright red or colored to indicate that it has reached a high temperature? Does the flame go out when removed from the source? What type of ash or residue, if any, is formed?

(6) Notice any odor given off by the fiber both while it is in the flame and after it is removed.

(7) Observe the ash or residue formed and what characteristics it has. Is it brittle? Bead shaped? Fluffy? The shape of the yarn? Or is there nearly no residue?

(8) Repeat for the filling yarn of woven fabrics.

(9) If the fabric does not have yarn structure, or if it is impossible to "deknit" a length of yarn from complex knitted structures, a small sliver of fabric can be cut and used in place of the yarn.

Microscopic evaluation

Microscopical examination is indispensable for positive identification of the several types of cellulosic and animal fibers, because the infrared spectra and solubilities will not distinguish between species. It is possible to be quite definitive in the identification of some fibers through microscopic observations. Fibers are mounted to provide views of both their lengthwise and crosswise dimensions. Unfortunately, several of the man-made fibers are similar in their microscopic appearance; therefore, additional analysis is required for specific identification. Photomicrographs of selected fibers may be used for comparison.

Solubility

The solubility of a fiber in specific chemical reagents is frequently a definitive means of specific fiber identification. Frequently, however, this process identifies only generic groups or categories. If the student combines the results of the burning test, the microscopic evaluation, and the chemical solubility test, it is possible, in many cases, to positively identify specific fibers.

In addition to aiding in fiber identification, familiarity with fiber behavior in various chemicals provides helpful information concerning the processing and care of textile fibers. A few substances used in such procedures as stain removal, cleaning, and laundering may damage some fibers. Obviously, knowing what chemicals to avoid in care would extend the life of a textile product.

Staining test

When fibers are white or off-white or when color can be stripped from fibers, staining techniques may be used as a part of identification. Specifically prepared mixtures of dyes are dissolved in water, or other chemicals if the specific stain mixture so specified, and the fibers are immersed in the solution and stained for a specified time and at a specified temperature. After staining, the fiber, yarn, or fabric pieces are dried and then compared with a known sample.

Instrumental analysis procedures

Certain man-made fibers are not easily identified by any of the testing procedures cited. Positive verification of some fibers depends on the use of one or more sophisticated instrumental techniques. These include testing for density, melting point, refractive index, index of birefringence, the use of X-ray diffraction machines, infrared spectrophotometers, chromatographs of various types, electron scanning microscopes, and polarizing microscopes. These are standard equipments in many university laboratories, testing laboratories, and research laboratories. Although they may not be available in departments where textile science is taught, they may be available in chemistry departments.

It should be noted most manufacturers offer a variety of fiber types of a specific generic class. Differences in tenacity, linear density, bulkiness, or the presence of inert delustrants normally do not interfere with analytic tests, but chemical modifications (for such purposes as increased dyeability with certain dyestuffs) may affect the infrared spectra and some of the physical properties, particularly the melting point. Many generic classes of fibers are sold with a variety of cross-section shapes designed for specific purposes. These differences will be evident upon microscopical examination of the fiber and may interfere with the measurements of refractive indices and birefringence.

Summary of test method

专业微课堂：
纤维鉴别

Generally, analyses by infrared spectroscopy and solubility relationships are the preferred methods for identifying man-made fibers. The analysis scheme based on solubility is very reliable. The infrared technique is a useful adjunct to the solubility test method. The other methods, especially microscopical examination are generally not suitable for positive identification of most man-made fibers and are useful primarily to support solubility and infrared spectra identifications.

(1) The fiber generic type is identified from its solubility in various reagents, using a solubility decision scheme (Fig.13.1).

(2) Alternatively, infrared spectra of fibers from textile materials to be identified are obtained using a FTIR (Fourier Transform Infrared) or a double-beam spectrophotometer. Identification of the fiber generic class is made by analysis of the fiber spectrum using a decision chart (Fig.13.2).

(3) For plant (native cellulose) and animal hair fibers microscopical examination of longitudinal and cross-sections is used to distinguish species.

(4) Additional physical properties of the fiber, such as density, melting point, regain, refractive indices, and birefringence are determined and are useful for confirming the identification (Table 13.1).

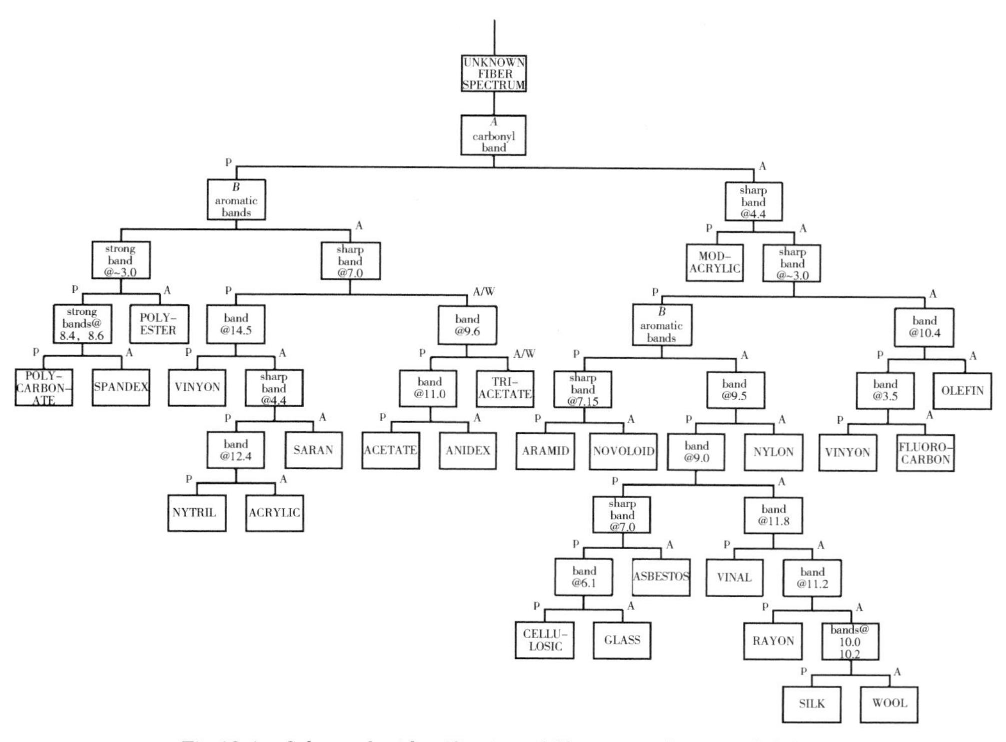

Fig.13.1 Scheme for identification of fibers according to solubility

(Source: ASTM Standard D276-12 Standard Test Methods for Identification of Fibers in Textiles. Note: Bands located according to wavelength in m. *A*—Acidify with excess HCl, add lead acetate dropwise. *B*—Rinse with water, allow to dry in room air. *C*-Some modarylic fibers cannot be distinguished from acrylic fibers in this solubility scheme. P = present, A/W = absent or weak)

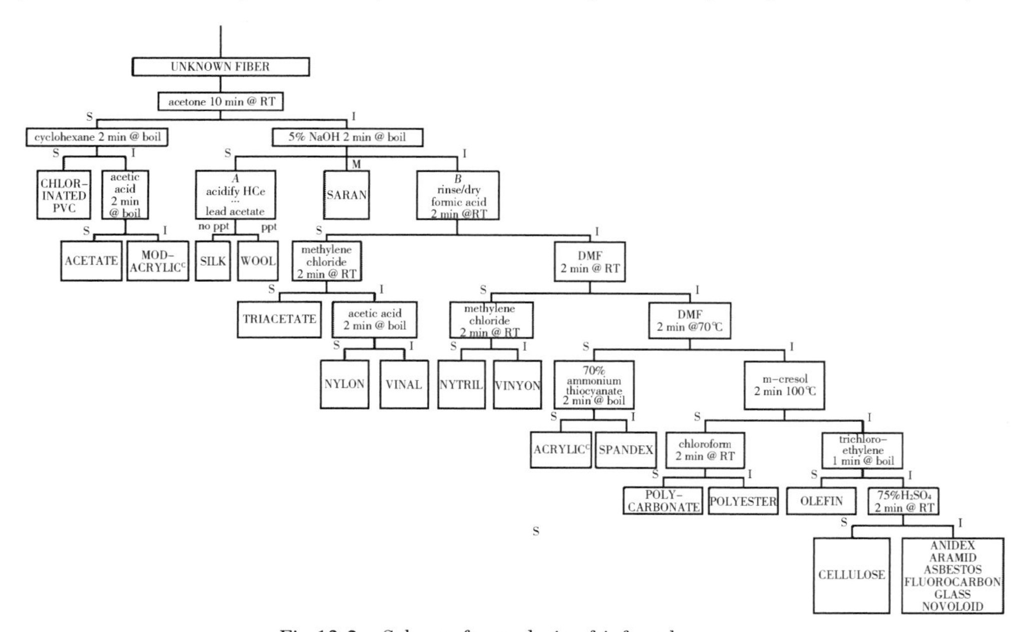

Fig.13.2 Scheme for analysis of infrared spectra

(Source: ASTM Standard D276-12 Standard Test Methods for Identification of Fibers in Textiles. Note: M = melts, S = soluble, I = insoluble, ppt = precipitate, RT = room temperature. *A*-located at; 5.75, *B*-located at 6.23, 6.30 & 6.70)

Table 13.1 Typical values of physical properties useful for identifying fibers

Fiber	Melting[A] Point/℃	Refractive Index[B]		Birefringence[B] ε-ω	Density/(mg/mm^3)
		Parallel to Fiber Axis, ε	Perpendicular to Fiber Axis, ω		
Acetate	260	1.479	1.477	0.002	1.32
Acrylic	dnm	1.524	1.520	0.004[C]	1.17
Anidex	s 190	[D]	[D]	[E]	1.22
Aramid					
Nomex®	371	1.790	1.662	0.128	1.37
Kevlar®	425	2.322	1.637	0.685	1.42
Asbestos	…	1.5-1.57	1.49	0.01-0.08	2.1-2.8
Cellulosic					
Flax	dnm	1.596	1.528	0.068	1.54
Cotton	dnm	1.580	1.533	0.047	1.54
Fluorocarbon	288	1.37	…	…	2.1
Glass	s 570	1.547	1.547	0.000	2.47-2.57
Modacrylc	dnm	1.536	1.531	0.005	1.28-1.37
Novoloid	dnm	1.650	1.648	0.002	1.29
Nylon					
nylon 6	219	1.568	1.515	0.053	1.14
nylon 6.6	254	1.582	1.519	0.063	1.14
Qiana®	275	1.554	1.510	0.044	1.03
nylon 4	265	1.550	…	…	1.25
nylon 11	185	1.55	1.51	0.04	1.04
Nytril	176	1.484	1.476	0.008	1.20
Olefin					
polyethylene	135	1.556	1.512	0.044	0.93
polyethylene	170	1.530	1.496	0.034	0.90
Polycarbonate	294	1.626	1.566	0.060	1.21
Polyester					
2 GT[E]	256	1.710	1.535	0.175[F]	1.38
4 GT[G]	227	1690	1.524	0.166	1.32
CHDM-T[H]	283	1.632	1.534	0.098	1.24
oxybenzoate[I]	225	1.662	1.568	0.094	1.34
Rayon					
cuprammonium	dnm	1.548	1.527	0.021	1.53
viscose	dnm	1.547	1.521	0.026	1.52

Continued

Fiber	Melting[A] Point/℃	Refractive Index[B]		Birefringence[B] ε-ω	Density/(mg/mm^3)
		Parallel to Fiber Axis, ε	Perpendicular to Fiber Axis, ω		
Saran	170	1.603	1.611	−0.008	1.62−1.75
Silk	dnm	1.591	1.538	0.053	1.35
Spandex	230	1.5	…	…	1.2
Triacetate	288	1.472	1.471	0.001	1.30
Vinal	dnm	1.543	1.513	0.030	1.30
Vinyon (PVC)	dnm	1.541	1.536	0.005[D]	1.40
Wool	dnm	1.556	1.547	0.009	1.31

A dnm indicates the fiber does not melt, *s* indcates softening point.

B The listed values are for specfic fibers which warrant the highly precise values given. For identification purposes these values should be regarded as indicating only the relative values of the properties.

C Varies, always weak, sometimes negative.

D The fiber is opaque.

E Ethylene glycol type.

F Staple and fully oriented filament yarns (FOY), partially oriented (POY) and undrawn yarns may have much lower values of birefringence and refractive index.

G 1, 4-butanediol type.

H 1, 4-cyclohexanedimethanol type.

I p-ethylene oxybenzoate type.

(Source: ASTM Standard D276-12 Standard Test Methods for Identification of Fibers in Textiles)

New Words and Expressions

identification 鉴别
methodology 方法学，方法论
qualitative 定性的
tentatively 试验性地
burning test 燃烧测试
warp 经纱
unravel 拆散
untwist 解捻，退捻
flammable fiber 可燃纤维，易燃纤维
forceps 镊子
odor 气味
residue 残渣
fluffy 毛茸茸的
solubility 溶解法，溶解
photomicrograph 显微照片
staining test 染色测试
melting point 熔点
refractive index 折射率
index of birefringence 双折射率
X-ray diffraction machine X 射线衍射仪
infrared spectrophotometer 红外分光光度计
chromatograph 色谱仪
polarizing microscope 偏振光显微镜
bulkiness 蓬松度
delustrants 消光剂
refractive indices 折射率

Notes to the Text

1. some tests yield insight into problems of processing and care of textile products：某些实验会发现纺织产品加工以及保养中的问题。句中的“yield”意思为“产生……的效果”，“insight”意思为“洞察”。
2. suggests interpretation：可译为“建议性的解释”。
3. ...which helps place a fiber into categories but seldom identifies a specific fiber：“place a fiber into categories”是指“将某纤维归为某一类”，“a specific fiber”指“某种具体的纤维”。
4. microscopic evaluation：显微结构评定。
5. ...which is a little more specific：更专业化一些。
6. to categorize fibers into generic groups：将纤维归到同类的组中。
7. the test will usually give the reaction of the fiber that burns most easily：指发生与容易燃烧纤维相关的反应。
8. approach the flame from its own level：在与火焰同高的位置接近火源。
9. Does the flame go out when removed from the source：“go out”意思为“熄灭”。
10. both while it is in the flame and after it is removed：在火焰中时和离开火焰时。
11. If the fabric does not have yarn structure：指一些毡类或起绒类织物。
12. deknit：这是一个自造的词，前缀 de-来自拉丁语，意为”away from”，这里表示从针织物上拆卸、拆散。
13. Fibers are mounted to provide views of both their lengthwise and crosswise dimensions：句中“are mounted”的意思是“安放”。
14. In addition to aiding in fiber identification：除了对纤维鉴别有帮助外。
15. or other chemicals if the specific stain mixture so specified：如果对特定的染色用混料有特殊要求，也可使用其他化学品。
16. electron scanning microscope：扫描电镜，一般写为“scanning electron microscope (SEM)”。
17. chemical modification：化学改性
18. Many generic classes of fibers are sold with a variety of cross-section shapes designed for specific purposes：许多同类纤维为了特定目的(如增强吸湿性等)而做成各种横截面形状进行销售。
19. the preferred methods for identifying man-made fibers：指鉴别人造纤维的首选方法。

Questions to the Text

1. How many methods can be used to identify textile fibers?
2. Describe the several steps in doing the burning test.
3. Tell your classmates about the microscopic observation test once you did.
4. What instruments can be used in fiber identification?

Lesson Fourteen

Staple Fiber Spinning

语言微课堂

Opening, cleaning, and blending

Staple fibers arrive at the yarn processing plant in large bales. To make yarns, fibers must be of similar length and relatively uniform so that the spun yarn can be of uniform quality. To accomplish this, fibers from a variety of production lots, fields, or animals must be blended together. Some fibers from each bale or carton are fed into the opener and blender.

Bales are placed into some type of automatic fiber feed unit. Metal fingers pull tufts of fiber from the bale. The fibers are separated and fed onto a spiked apron or lattice that carries the fibers from the feed area to the opening and cleaning area. The opening operation separates the fibers into a loose, fluffy mass. It is important to separate or "open" the fiber mass to a single fiber state, or as close to that as possible.

These loose fibers are fed into a hopper, where a measured amount is laid on a conveyor belt and delivered to the picking unit, where additional blending occurs. The picker further opens, cleans, and blends fibers through a system of rollers and forced air.

During this operation most of the dirt and impurities that might be present are removed by either gravity or centrifugal force. Cotton fibers receive more opening and blending than man-made fibers, since they have more impurities and greater variation than do man-made fibers.

In the intermittent system the blend of fibers is blown onto a collecting cylinder to form a fiber layer. As the cylinder rotates, the layer of fibers is rolled off to form the picker lap which is placed at the rear of the card frame to supply fibers. In the automatic system the fibers are held in a hopper and fed in a loose form directly to the card.

The continuous system for making ring-spun yarns takes fibers directly from the bale and processes them automatically through to at least the card sliver state. Further steps may be connected to the card so that the card sliver moves on through additional processes automatically. Slivers from the card go directly to the drawing frames and may actually be fed continuously through to the roving process and spinning process.

Yarns made on automatic equipment tend to be more uniform and may be stronger than discontinuous process yarn. Production speed is considerably faster for continuous processes, labor costs are reduced, and plants stay cleaner.

Carding

Carding continues the cleaning of the fibers; it removes fibers too short for use in yarns. The process partially aligns the fibers so that their longitudinal axes are somewhat parallel. Carding is accomplished by wire cards. Wire cards contain two layers of card clothing consisting of wire flats (rectangular shapes) in which fine wire pins are anchored. The flats are attached to a steel cylinder and to an endless belt that rotates over the top portion of the cylinder. The two sets of pins move in the same direction, but at different speeds, to tease the fibers into a filmy layer, so that a thin web of fibers is formed on the cylinder. This thin web is gathered into a soft mass and pulled into a ropelike strand of fibers, called a sliver. The sliver is pulled through a cone-shaped outlet and doffed or delivered to cans or to a conveyer belt. The card sliver is not completely uniform in diameter, and the fibers are somewhat random in arrangement.

Combing

Some fabrics are made of yarns that have received only the carding operation prior to drawing and roving formation. However, some fabrics require finer quality yarns of superior evenness, smoothness, fineness, and strength, particularly high-quality cotton fabrics, and these require combed yarns rather than carded. When cotton and man-made fibers are combined, it is common for the cotton fibers to receive the combing step before the two types of fibers are combined. For yarns that require the additional step, the card sliver goes through the breaker-drawing step and then to the combing operation.

Card slivers are fed to the breaker-drawing frame, where several card slivers are combined. The break-drawing unit pulls out the fibers into a thin layer, and reforms a new sliver. The drawing is accomplished by controlling the speed of a series of rollers, each set of which operates faster than the one behind it. The layer of slivers is pulled through the rollers at increasing speeds; as the layer leaves the unit, the fibers are pulled into a new sliver and delivered to cans ready for the combing frame.

Forty-eight slivers from the breaker-drawing unit are combined to form the lap for the comber. The slivers are fed through a lapper that makes them to a thin layer of fibers that is wound onto a roll. These rolls, each weighing about 13.6kg, are taken to the combing frame.

The layer of fibers is fed into the combing area, where fine metal wires clean out remaining short fibers and impurities and further parallel the fiber in the comber lap. During the combing operation as much as 20% of the fibers may be removed. This waste is sold to manufacturers of nonwoven products and to others who have use for short fibers. The fibers remained form a thin web or layer; this web of fibers is pulled together, fed through a cone that helps to hold the fibers together, and delivered as a comb sliver.

Finisher drawing

Slivers from either the carding unit or the combing unit, depending on the ultimate yarn desired, are processed through the finisher-drawing or drafting frame. This is the process by which fibers of different types can be blended together to form blended yarns. Eight slivers are drawn together to produce the drawn sliver. If a 50/50 polyester/cotton blend is to be made, there will be four slivers of polyester fiber and four of cotton; if a 65/35 blend is ordered, there will be five polyester slivers and three cotton. As with breaker drawing, rollers moving at different speeds smooth and combine the slivers and pull them into a thin layer and then into the drawn sliver.

The finisher-drawing operation is usually repeated. Eight slivers from the first finisher-drawing step are subjected to the same operation. The drawn sliver is about the same size as, or perhaps slightly smaller than, the card or comb slivers. As yet no twist has been imparted into the fiber assemblage, although the delivery of the sliver to the cans tends to twist the sliver slightly.

Roving

Slivers from the finisher drawing are taken to the roving frame, where each sliver will be attenuated until it measures approximately one-eighth of its original diameter. The drawn sliver is fed between sets of rollers. Each set of rollers rotates faster than the set behind it, the front rollers rotate about ten times faster than the back set of rollers. This pulls the fibers out, reduces the diameter of the strand, and further parallels the fibers. A slight amount of twist is imparted to give strength. The new strand, called roving, is laid onto a bobbin. Roving is wound onto the bobbin package at approximately 27 meters per minute. The full bobbins are doffed from the frame and delivered to the spinning frame.

专业微课堂：短纤纺

Spinning

专业微课堂：环锭细纱机

The final process in ring spinning of single yarn is the spinning operation which is shown in Fig. 14. 1. During spinning the roving is attenuated to the desired diameter, called the final draft, and the desired amount of twist is inserted. The roving is fed down into the spinning area, where it feeds between sets of rollers. Just as in the roving step, the front set of rollers rotates faster than the back set. In spinning, the front set rotates about 30 times faster than the back set. This difference in speed attenuates the yarn and makes it even, smooth, and uniform. The attenuated yarn is fed down, guided through a U-shaped guide on a ring, called a traveler, which moves around the take-up package or bobbin, hence the name "ring spinning". The movement of the traveler and the turning of the spindle on which the bobbin is held combine to introduce wind into the yarn. The spindle turns at about 13, 000 revolutions per minute; the traveler is slightly slower. Approximately 11 meters of yarn is wound onto the bobbin per minute. The size of yarn and the amount of turns, cited as turns

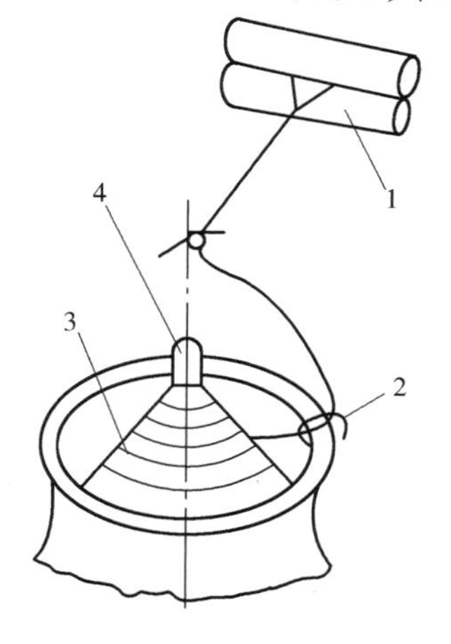

Fig.14.1 Ring spinning
1—Front roller 2—Traveler
3—Bobbin 4—Spindle

per meter, can be controlled. The yarn manufacturing steps discussed to this point produce a single or singles yarn.

New Words and Expressions

carton 纸板箱
opener 开棉机,开松机
blender 混棉机,混合机
spike 置凸钉
apron 胶圈,皮板输送带
lattice 输送帘子
opening 开松
cleaning 除杂
hopper 料斗,棉箱
conveyor belt 传送带
picking unit 清棉装置
forced air 高压气流
intermittent system 间歇式纺纱系统
picker 清棉机
picker lap 清棉棉卷
ring-spun yarn 环锭纱
drawing frame 并条机
wire card 钢丝梳棉机
card clothing 针布
wire flat 盖板
tease 梳理
cylinder 锡林
doff 落卷,落筒
can 条筒
combed yarn 精梳纱
carded 粗梳的
breaker-drawing 预并条机
roller 罗拉
lapper 成网机,成网机构
comber 精梳机
finisher-drawing 并条机
card unit 梳理机构
assemblage 集合体
roving frame 粗纱机
attenuate 使变细
bobbin 筒管
spinning frame 细纱机
ring spinning 环锭纺纱
draft 牵伸
traveler 钢丝圈
ring 钢领
spindle 锭子
single 单纱

Notes to the Text

1. production lots:生产的批次。
2. since they have more impurities and greater variation than do man-made fibers:句中的"variation"指纤维具有不同的长度,卷曲,状态等。
3. the layer of fibers is rolled off:纤维层被输送出来。
4. The continuous system for making ring-spun yarns takes fibers directly from the bale and processes them automatically through to at least the card sliver state:连续式环锭纺纱系统从棉包开始,连续自动加工至少到生条阶段。
5. Some fabrics are made of yarns that have received only this carding operation prior to drawing and roving formation:有的织物使用的纱线,在加工过程中仅经过一次梳理,然后是并条、粗纱工序。

6. fibers are combed as well as carded：纤维不仅要梳理，还要经过精梳。
7. rollers moving at different speeds smooth and combine the slivers and：“smooth”可译为“理顺”。

Questions to the Text

1. Why fiber blending is necessary?
2. Describe the opening, cleaning and blending process.
3. How does the carding frame work?
4. What is a carded yarn?
5. What is the purpose of the combing process?
6. Where does the waste in the combing frame go?
7. In order to make blended yarn, what should we do in the finisher-drawing step?
8. How does a roving frame work?
9. Describe the ring spinning process.
10. Why front rollers run faster than the back rollers in the spinning frame?

Lesson Fifteen

New Spinning System (1)

Rotor spinning

Rotor spinning belongs to open-end spinning. "Open-end" should be applied only to systems where a break occurs in the fiber system. Break spinning and open-end spinning are synonymous. The essential steps in open-end spinning include the following. A coarse sliver of fibers is fed to an opening system, which opens the sliver to the point where the fibers are individual entities; the individual fibers are fed forward; they are collected together on a small surface and pulled from the surface as a thin layer constantly adding to the open end or tail of the forming yarn. The thin layer is attenuated, twist is inserted, and the resulting yarn is wound onto bobbins.

Rotor spinning involves the contact of fibers in or on rotating devices such as funnels, cones, sieves, or needled surfaces. A sliver of fibers is fed into the unit, a current of air forces the fibers into a loose form, and they are collected on the rotating device. As they collect, they are pulled off the rotor by mechanical means to form the yarn, and the new yarn is wound onto packages. Twist is inserted as the yarn is removed from the rotor.

Rotor spinning is a popular process and has replaced some ring-spinning units. There are some differences in quality and characteristics between ring-spinning and rotor-spun yarns, but both have some advantages. Many of the disadvantages associated with early rotor-spun yarns are disappearing as the process is improved and perfected. Important aspects of rotor-spun yarn include the facts that they require considerably less plant space, energy requirements and labor needs are reduced, operation is cleaner, and production of yarn is considerably faster than for ring spinning. The same amount and size of yarn spun on the rotor machine in 0.02 hour would take 0.3 hour on a ring spinning frame.

Early rotor-spun yarns tended to be coarse, but modern rotor-spinning machines produce fine yarns as well as medium to heavy or coarse yarns. Furthermore, yarns are even and defects are reduced. Fiber defects may cause problems, but newer systems can eliminate such fibers and discharge them as waste prior to yarn formation.

Properties of rotor-spun yarns compare favorably with those of ring-spun. Fiber distribution is more uniform in rotor-spun than ring-spun, and fiber migration is less in rotor-spun than in ring-spun yarns. Because of the way fibers arrange themselves in rotor-spun yarns, dyeing behavior may differ from that of ring-spun yarns. It has been stated that deeper colors and differential dye techniques are possible with rotor-spun yarns.

End uses for rotor-spun yarns include apparel, home furnishings, and industrial fabrics. In fact, with the current state of the art of spinning, open-end spun yarns, particularly rotor-spun, compete with ring-spun in nearly every possible end use.

Vortex spinning

The sliver is fed to 4-over-4 (or a four-pair) drafting unit. As the fibers come out of the front rollers, they are sucked into the spiral-shaped opening of the air jet nozzle which is shown in Fig.1. The nozzle provides a swirling air current which twists the fibers. A guide needle within the nozzle controls the movement of the fibres towards a hollow spindle. After the fibers have passed through the nozzle, they twine over the hollow spindle. The leading ends of the fiber bundle are drawn into the hollow spindle by the fibres of the preceding portion of the fiber bundle being twisted into a spun yarn. The finished yarn is then wound onto a package. The schematic drawing of the principle of vortex spinning is shown in Fig.15.1.

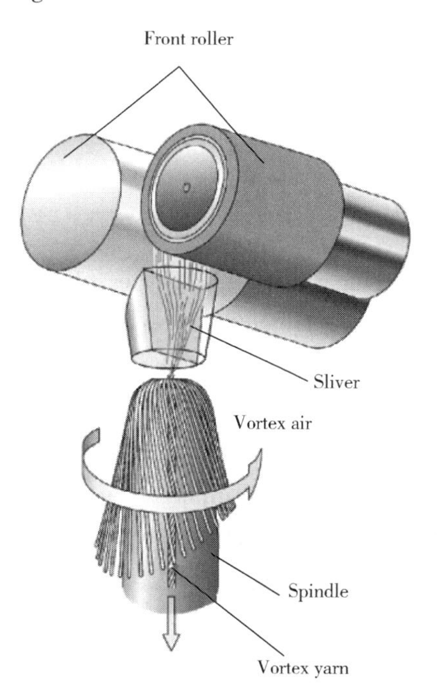

Fig.15.1 Principle of vortex spinning

Compact spinning

In conventional ring spinning, fibers in the selvedge of strand emerging from front roller nip do not get fully integrated into the yarn because of the restriction to twist flow by the spinning triangle. These fibers show up partly as protruding hairs or as wild fibers. The hairiness gives a rough feel to the yarn. Long protruding hairs from the yarn contribute to multiple breaks in weaving and fabric faults.

The spinning triangle exists because of higher width of the strand as compared to final yarn diameter which is shown in Fig.15.2. Further the fibers are tensioned to varying extent depending upon their position in the spinning triangle. As a result full realization of fibers trength is not achieved in the yarn. This problem is solved by applying the compact spinning systems that increases yarn quality. It is carried out by means of narrowing and decreasing the width of the band of fibers which come out from the drawing apparatus before it is twisted into yarn, and by the elimination of the spinning triangle.

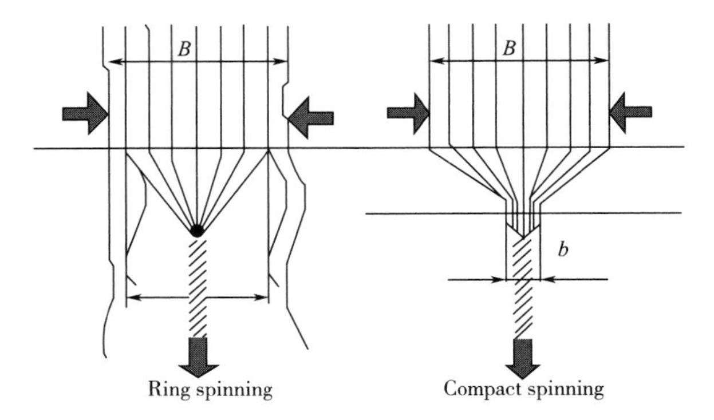

Fig.15.2 Spinning triangles in ring and compact spinning

B—the width of the band of fibers *b*— the width of the band of fibers

The spinning triangle is the critical weak spot of the spinning process. The spinning triangle prevents the edge fibers from being completely incorporated into the yarn body. However, in compact spinning, the drafted fibers emerging from the nip line of the front roller of the drafting arrangement are condensed in a line. This condensation happens in so called "Condensing Zone" following the main drafting zone.

Different machine manufactures are using different methods to condense the fibers emerging out from the front roller. These methods are: ①Aerodynamic compacting system: a) Suction by drum and b) Suction through perforated apron. ②Mechanical compacting system. ③Magnetic compacting system. In the aerodynamical compacting system the condensation of the fibers strand take place with help of perforated drum or apron. The examples of aerodynamical compacting system are Com4Spin® of Rieter, Elite® Compact Spinning by Suessen, CompACT3 by Zinser, Com4®wool by Cognetex, Olfil system by Marzoli, Toyota's compact spinning, etc.

Siro spinning

Two rovings are passed individually through a slightly modified, but generally conventional drafting arrangement of a normal ring spinning machine. The fiber strands, attenuated by a draft in the normal range, leave the delivery roller separately. At this point, they are each subjected to twist generated by a common spindle; thus, within the spinning triangle, they are twisted into two single yarns, and these are simultaneously bound together to form a composite yarn. Each of the two single strands and the resulting composite yarn contains twist, and the direction of twist is the same for both the single ends and the composite product.

This spinning process primarily offers economic advantages, because the production of the ring spinning and winding machines is roughly doubled (two ends instead of one at approximately the same speed). In addition, plying and twisting are eliminated. However, due to the different twist structure, it cannot completely replace the conventional 2-fold yarn process.

New Words and Expressions

open-end spinning 自由端纺纱，气流纺纱
synonym 同义词
break spinning 自由端纺纱，气流纺纱
rotor spinning 气流纺纱
sieve 筛网
rotor 纺纱杯
warping 整经
fiber migration 纤维迁移
differential dye technique 差异染色技术
compare favourably with 似优于，不亚于
the spiral-shaped opening 螺旋形通道
nozzle 喷嘴
A guide needle 引导针
a hollow spindle 空心锭子
selvedge 边缘
protruding 突出的
the spinning triangle 纺纱三角区
perforated apron 带孔的胶圈
drum 集聚罗拉
perforated drum 集聚罗拉

Notes to the Text

1. which opens the sliver to the point where the fibers are individual entities：将条子开松直至纤维呈独立存在状态。
2. pieces slivers：将条子连接。
3. compare favorably：有优势。
4. with the current state of the art of spinning：句中“art”可译为“技术”。
5. The leading ends of the fiber bundle are drawn into the hollow spindle：句中“leading”可译为“头端”。
6. It is carried out by means of narrowing and decreasing the width of the band of fibers：句中“narrowing and decreasing”可译为“减小或消除”。
7. This condensation happens in so called “Condensing Zone” following the main drafting zone：句中“Condensing Zone”可译为“集聚区”。
8. The examples of aerodynamical compacting system are Com4Spin® of Rieter, Elite® Compact Spinning by Suessen, CompACT3 by Zinser, etc：通过空气负压实现紧密纺的有瑞士立达的卡摩纺、绪森倚丽的紧密纺、青泽的 CompACT3 紧密纺等。

Questions to the Text

1. Describe the open-end spinning process.
2. What are the advantages of the rotor spinning technique?

3. Say something about the properties of the rotor-spun yarn.
4. Please introduce the process of vortex spinning.
5. Why to develop the compact spinning system?
6. What is the principle for compact spinning system?
7. What methods can be adopted to condense the fibers emerging from the front roller?
8. What is the process of Siro spinning system?

Lesson Sixteen

New Spinning System (2) *

Friction-spun yarns

A modification of existing open-end technique has been introduced that combines mechanical and air current operations. The original system was called DREF after the developer, Dr. Fueher. Several modifications have been made in this technique, and it is now generally referred to as friction spinning.

In this process, the sliver enters the system and the fibers are separated and spread onto a combing or carding roll as shown in Fig.16.1. They are doffed from this roll and transported by air to the friction zone. Two cylinders rotating in the same direction pull the fibers together to form the yarn.

Friction-spun yarns are uniform, quite free of impurities, fuller with more body than ring-spun yarns, but they tend to have less strength. These yarns can be used successfully for apparel fabrics, home furnishing fabrics, and industrial fabrics of certain types.

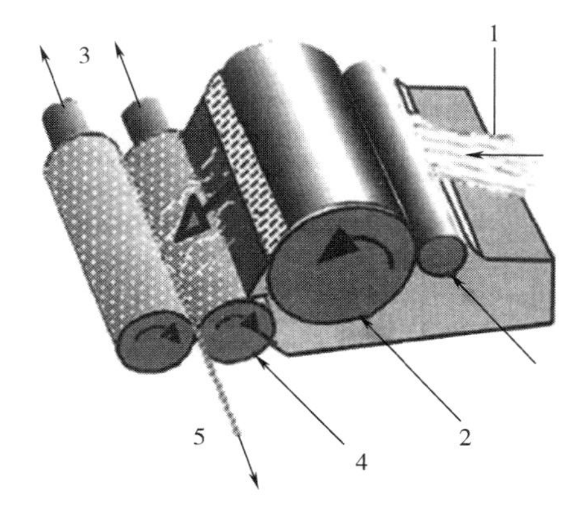

Fig.16.1 Friction spinning

1—sliver 2—combing or carding roll 3—air 4—two friction rolls 5—yarn

Self-twisted yarns

Self-twisted yarns are essentially two-strand or two-ply (or more) yarns. They may involve one ply or strand of staple fibers and one of filament, two strands of staple fibers, or two of filaments, or multiple combinations of these. A common type is the Selfil yarn, which is composed of one strand of staple fibers with some twist and a filament strand wrapped around the staple strand. The filament strand alternates S and Z twist. A number of combinations of these strands have been made.

Self-twist spinning produces yarns more cheaply than ring spinning and considerably faster with less labor. Selfil yarns are slightly more even than ring-spun and are much stronger because of the filament wrapping. Furthermore, in knitted structures, Selfil yarns snag less than ring-spun yarns.

Twistless yarns

Fibers are held together with adhesive to form yarn in the twistless process. A roving is made and then drawn to a fine strand; during the drawing an adhesive is applied by rollers. A common adhesive is starch. The drawn yarn, coated with adhesive, is wound onto packages; the package is steamed, and the adhesive bonds the fibers together.

Characteristics of twistless yarns as compared with ring-spun yarns include the following. Twistless yarn is stiffer than ring-spun yarn and has greater luster, less elongation, and better covering power owing to its ribbonlike character. Also, it is more uniform and somewhat weaker than ring-spun yarn.

Core-spun yarn

专业微课堂：
包芯纱

Core-spun yarns have a central core with a second layer or sheath of fibers wrapped around it. Both core and wrapping may be the same fiber type, or two different fiber types (or more) may be used. Further, one part may be of filament fibers while the other is of staple fibers; or both parts may be of either filament or staple-length fibers. The outer sheath may completely cover the core. When the outer sheath does not hide the center completely, the two components create interesting appearance effects.

One of the most common types of core-spun yarns is that involving a core of a stretch filament such as spandex with a covering of staple fibers. It may also be used for a variety of other fiber types and end uses.

A relatively new process for making a special type of yarn is called coverspun. It can use any type of staple fibers, but it is especially desirable for wool, and it uses any type of filament fibers. Coverspun is described as a bicomponent yarn made of a center or core of staple fibers that are wrapped with filament fibers that serve as the binder. The process uses a conventional roving as the supply source for the staple fiber core.

A modification of the process involves three components, a core of filament fibers encircled by staple fibers, then wrapped by filament fibers.

An important factor of this process is that the staple fibers are not twisted, but rather are bundled together. This means yarns can be made rapidly. Coverspun yarns have the following properties. They are generally stronger than ring-spun yarns. This is due to the filament wrapping and to a high interfiber friction. They are more even than ring-spun yarns. They have an elongation equal to ring-spun yarns. They can be used equally well in knitted or woven fabrics. Fabrics made of these yarns do not look as if filaments have been used. Covering power of the yarn is good. Pilling of yarns in fabrics is reduced. Fabrics are soft and supple and can be light in weight and thin with little bulk. Yarns have an attractive appearance.

New Words and Expressions

friction spinning　摩擦纺纱
DREF　德雷夫纺纱法，尘笼纺
body　身骨
self-twisted yarn　自捻纱
Selfil yarn　加长丝自捻纱
twistless yarn　无捻纱
core-spun yarn　包芯纱
spandex　斯潘德克斯弹性纤维，氨纶
coverspun　包绕纺纱
bicomponent　双组分
supple　柔软的

Notes to the Text

1. The original system was called DREF after the developer：“after the developer”意思是“按照发明者命名”。
2. drafts the sliver into a core of fibers：将纤维条子牵伸成芯纤维。
3. the package is steamed：卷装被烘干。
4. interfiber friction：纤维间的摩擦。

Questions to the Text

1. How is the friction spinning yarn processed?
2. What are the differences between the friction spinning yarn and the ring-spun yarn?
3. What is the structure of the self-twisted yarn?
4. What are the major properties of the self-twisted yarn?
5. How is the twistless yarn processed?
6. What is a core-spun yarn?
7. Describe the characteristics of the coverspun yarn.

语言微课堂

Lesson Seventeen

Relationship Between Yarn Structure and Fabric Performance

Yarn construction can alter the properties of the fabrics that are made from them. The cloth properties that are most dependent upon yarn construction are those affecting appearance, comfort, durability and maintenance.

The amount of light reflected from the surface of the cloth, and the way in which it is reflected, defines the property we call luster. Although cloth construction can affect the way light is reflected, it is the yarns and fibers that have the greatest effect. A highly reflecting fiber, such as silk or nylon, appears very shiny. A highly crimped fiber, like wool, has a duller appearance. The amount of twist in the yarn also modifies the way light is reflected. If there is no twist in the yarn, the light is scattered from each individual fiber and the yarn appears dull. With a small amount of twist, the fibers are aligned and light is scattered from their sides; this increases the amount of light reflected to the eye of the observer. With increasing twist, the yarn becomes smoother, and the amount of reflected light increases. However, as the yarns are overtwisted, ridges appear on the surface of the yarn. These ridges tend to scatter the light away from the observer, with a concomitant loss of luster.

The warmth of a garment is determined by the insulating capacity of the cloth from which it is made. The insulating properties of cloth are derived from the trapped air retained between the fibers of the yarns. Thus, the ability of the cloth to retain body heat depends upon the tightness of the construction and the bulk of the yarns. A loose, open construction allows air to pass between the yarns. This flow of air enhances thermal interchange between the outside air and the air held between the garment and the body. Such a material will feel cool. A tight, compact construction prevents air from passing between the yarns, and feels warmer. The yarn properties, however, can reverse the behavior. Consider a tweed jacket made of a low yarn count cloth woven from bulky wool yarn. Compare it to a cotton blouse or shirt of a higher yarn count. The wool jacket will be much warmer because the bulky yarns trap air between the fibers. The trapped air warms up and, in the absence of wind, acts as an insulating layer between the body and the atmosphere. The cotton yarns are less bulky. They cannot trap as much air as the wool yarns, and do not maintain an effective barrier against heat transfer.

思政微课堂：纱线结构与性能

The moisture permeability of a garment is affected in much the same way with the insulating properties. We all have a thin film of water covering our skin. It is the evaporation of this surface water that allows us to regulate our body temperature. As the water evaporates into the atmosphere, heat is removed from the body.

If the surrounding air is dry, or if there is a breeze, the moisture evaporates readily and we feel cool. If the atmosphere is humid and the air is still, our body moisture is not readily removed, and we feel warm. On very sultry days we feel clammy and uncomfortable. If the cloth is compact and in close contact with the skin, it is difficult for moisture to penetrate the cloth. For example, filament yarns are very smooth. Cloth made from them tends to lie flat on the body. The moisture permeability will be low. Spun yarns, however, have stray fibers projecting from their surface. These stray fibers help to support the cloth and reduce its contact with the skin. This helps to improve moisture permeability and make the garment feel more comfortable. Today' s filament yarns are often textured in an attempt to approximate the hairiness of the spun yarns and make fabrics more comfortable.

The durability of a cloth is strongly dependent upon the yarns used in its construction. Filament yarns usually provide very high strength and abrasion resistance. Their strength derives from the fact that all the fibers are the same length, thus sharing equally in the load placed upon the yarn. The abrasion resistance of the filament yarns is often due to the high strength of the fibers. However, filament yarns of weak material, if given a moderate amount of twist, can withstand the cutting motion of abrasive forces acting across the fibers.

A major drawback of the filament yarns is their tendency to snag and pill. If one of the fibers of a filament yarn is broken, the lost end remains attached to the yarn. It cannot be pulled out because it is too strong. Often it is too strong to simply break away. This fiber curls itself up into a little ball, called a pill, which remains on the surface of the cloth. If enough pills form, the material becomes unsightly. Modification of the fibers has made current filament yarns much less likely to pill than the early man-made products.

The ease of maintenance of a cloth can be modified by the types of yarns used in its construction. Loosely twisted, bulky yarns are not able to resist penetration by soil or staining agents. Tight yarns with a high level of twist readily shed soil. In addition, the more open yarns are more easily compacted by the mechanical action of laundering. This increases the propensity of the fabric to shrink. In general, smooth, highly twisted yarns give better maintenance characteristics to a material.

The strength of a spun yarn follows a curve similar to that for the luster of a spun yarn, but for different reasons. A yarn with no twist can have no strength; there are no forces to hold the fibers together. As the yarn is twisted, fiber-to-fiber friction is increased and the yarn becomes stronger.

The abrasion resistance of spun yarns follows a pattern similar to that for strength. As the level of twist increases, the yarn becomes more compact. Abrasive stress is distributed over a number of fibers rather than a few, and it is more difficult to break or tear individual fibers. If the yarns are overtwisted, the fibers are subject to large internal stresses and will be easily broken by the application of outside forces.

The texture of a fabric can affect our feeling of comfort. If it feels smooth and silky, it is good; if it feels slick and waxy, it is bad. Fuzzy and warm is fine for winter pajamas, but hairy and coarse is uncomfortable. The texture of the yarn must be chosen to match the end use of the cloth. The fuzzy, comfortable feel of wool or textured acrylic is well suited to fall-or winter-weight clothing. The fine, highly twisted yarns of combed cotton or linen yield a smooth pleasant surface that is desirable for summer wear. The same effect is produced by smooth filament of rayon or acetate. It would be foolish to use hairy yarns in a satin construction, since satins are intended to provide a smooth surface that highlights the luster of the yarns. In uses such as linings, it is meant to ease dressing or disrobing. If the yarns are hairy, the surface fibers will diffuse the reflection of light and the fabric will lose its luster. In the second example, the increased friction caused by the protruding fibers will counteract the purpose of the satin weave.

New Words and Expressions

counteract 抵消，中和，阻碍
stray v. 迷路，偏离，漂泊，漂泊游荡；
adj. 迷路的，离群的，偶遇的
crimp 卷曲
scatter 消散
overtwist 加捻过度
ridge 凸棱
bulk 蓬松
tweed 粗花呢
jacket 夹克
bulky 蓬松的
blouse 女衬衫，罩衫
barrier 阻挡层
hairiness 毛羽
snag 钩丝
unsightly 不雅观的
shed soil 阻挡尘土
texture 质地
slick 滑溜溜的
fuzzy 毛茸茸的
pajamas 睡衣
acetate 醋酯
satin 经面缎纹
lining 衣服衬里
disrobe 脱衣
diffuse 扩散，漫射

Notes to the Text

1. open construction：稀疏的结构；稀疏的织物则为：open fabric。
2. The moisture permeability of a garment is affected in much the same way with the insulating properties：指影响透湿性和隔热性的因素相类似。
3. stray fibers：指纤维的线头。
4. modification of the fibers：纤维改性。
5. The strength of a spun yarn follows a curve similar to that for the luster of a spun yarn：指纱线捻度对强度和光泽的影响具有相同的规律。句中的“curve”指相关的规律曲线。
6. yarns of combed cotton：精梳棉纱。

Questions to the Text

1. What fabric properties can be affected by the yarn structure?

2. How does the yarn construction influence the luster of the cloth?
3. What kinds of yarns are suitable for winter garment?
4. What kinds of yarn structure are beneficial for ease maintenance?
5. Why filament can provide a higher strength in the fabric?
6. Describe the pill forming process.
7. How to choose the yarn to match the end use as far as fabric texture is concerned?

Lesson Eighteen

Yarn Properties(1)

语言微课堂

The assessment of yarn properties has been the subject of a considerable amount of work. There are many characteristics which combine to give an individual yarn its own distinctive properties.

Linear density

A system of denoting the fineness of a yarn by weighing a known length of it has evolved. This quantity is known as the linear density and it can be measured with a high degree of accuracy if a sufficient length of yarn is used.

The yarn number or yarn count is still in use today (Note that "yarn count" also designates the number of ends and picks in a woven cloth. Be aware of this overlap in nomenclature and don't confuse the two applications). Yarn number is defined as the number of standard hanks per pound of yarn. The standard hank varies with the fiber. Cotton and spun-silk yarns are measured by an 840yd (756m) hank; worsted (a fine wool yarn) is measured by the 560yd (504m) hank; linen and standard wool yarns are measured by the 300yd (270m) hank. The lengths of the standard hanks were chosen as convenient values for the yarn sizes in use at the time the system was devised. By this system, a no.50 cotton is a fine yarn in which 50 hanks, each 840 yards long, weighs one pound. Note that each hank weighs only 1/50 pound. A no.10 cotton requires 10 hanks to make up one pound, and is much coarser. A no.1 cotton is a very coarse yarn. Note that the higher the yarn number, the finer the yarn.

Denier is defined as the weight, in grams, of 9,000 meters of filament or yarn. Thus a filament yarn is characterized by its denier and the number of fibers in the yarn. For example, an untwisted 300-denier yarn containing 50 filaments is labeled 300/50/0. With a 100 turns per meter(tpm) twist in the S direction, it would be labeled as 300/50/100S. Note that this nomenclature allows one to calculate the size of each filament in the yarn. In the example mentioned, each of the filaments is 300/50 = 6 denier.

Tex is defined as the weight in grams of 1,000 meters of yarn, whether spun or filament. It is expected that the tex system will replace other methods of defining yarns in the near future.

Yarn twist

Twisting is a mechanism used to consolidate fibers into yarns and provide yarn strength. Yarn twist is typically described using two parameters: twist level and twist direction. Twist level is commonly expressed by the number of turns of twist per unit length. Another way to express twist level is by using the twist multiplier.

In spun yarns, fibers may be twisted in clockwise or counterclockwise directions. Commonly, these are described as Z twist or S twist (Fig.18.1). Yarns, cords, and cables may be twisted in either S or Z direction. An S-twist yarn is one in which the fibers follow a spiral pattern parallel to the center bar of the letter S. In a Z-twist yarn the fibers are parallel to the center bar of the letter Z. Often yarns are too fine to allow the unaided eyes to discern the twist direction. However, if you hold a short length of yarn vertically in your left hand and rotate it between the thumb and forefinger of your right hand while pulling upward, you can determine the twist direction. When it is rolled clockwise by your right hand, an S-twist yarn will become looser and a Z-twist yarn tighter. The direction of the twist does not affect the yarn's performance. However, it is customary to produce cotton and linen yarns with a Z-twist and woolen and worsted yarns with an S-twist. This custom has its roots in medieval consumer protection laws; consumers could easily find out whether a cloth was really wool by simply untwisting one of the yarns and checking the direction of the twist.

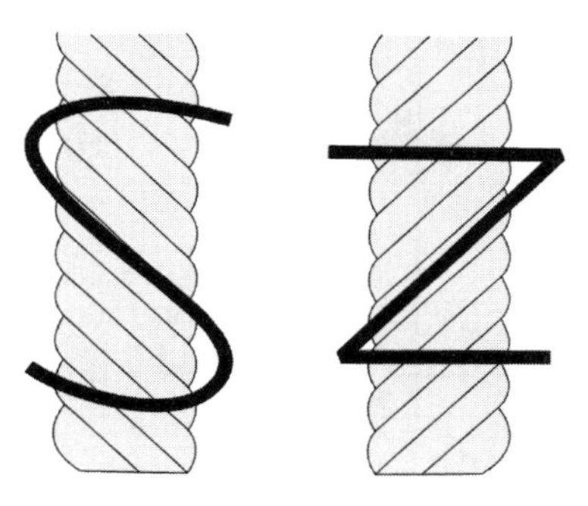

Fig.18.1 S-, Z-twist direction in yarns

专业微课堂：纱线捻度和线密度

思政微课堂：严灏景教授

As the yarn is twisted, internal forces are built up in the direction of the twist. These forces are known as torque. In general, each turn adds the same amount of torque to the yarn, so yarns with the same number of turns per meter have about the same amount of torque. Since the torque is approximately proportional to the tpm, the formation of balanced and unbalanced yarns may be illustrated by counting the turns per meter. Start with a pair of single yarns having 200 tpm in the S direction. Twist them together with another 100 tpm in the S direction. The internal torque on each of the singles will be about 300 tpm, while the torque on the two-ply yarn will be 100 in the S direction. This yarn will be unbalanced, and will curl. If the same two singles are plied into a double by twisting 100 tpm in the Z direction, the counterclockwise twist will tend to unwrap the single yarns and the internal torque of the single yarns will be reduced to 100 in the S direction. The external torque will be 100 in the Z direction. The internal and external torques balance each other and the yarn will be balanced. It is important to the manufacturer that the yarn be balanced during cloth construction because the kinks and curls that may be formed by unbalanced yarns will become entangled in the machinery and cause imperfections in the product. This simple analysis holds only for two-ply yarns. More complicated structures, such as cords and cables, require a more sophisticated approach.

Strength and elongation

Breaking strength is the maximum tensile force recorded in extending a test piece to breaking point as shown in Fig. 18.2. Stress is a way of expressing the force on a material in a way that allows for the effect of the cross-sectional area of the specimen on the force needed to break it. For many cases the cross-sectional area is not clearly defined. Therefore stress is only used in a limited number of applications involving fibers.

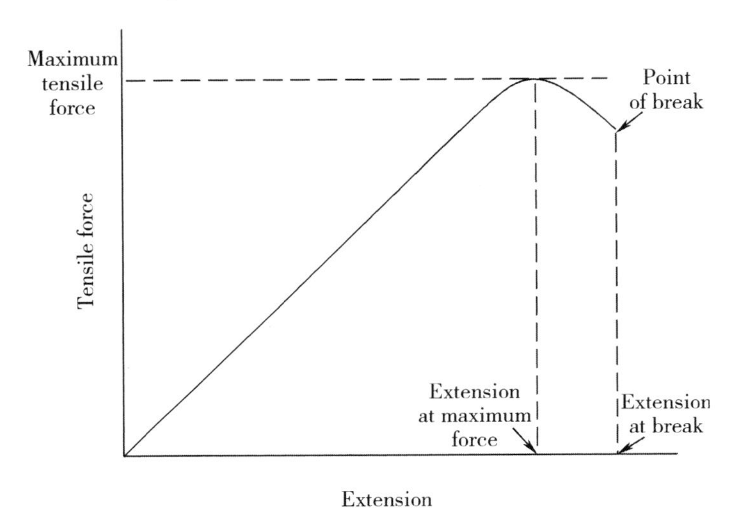

Fig. 18.2 A force extension curve

$$\text{Stress} = \frac{\text{force applied}}{\text{cross-sectional area}}$$

Specific stress is a more useful measurement of stress in the case of yarns as their cross-sectional area is not known. The linear density of the yarn is used instead of the cross-sectional area as a measure of yarn thickness. It is defined as the ratio of force to the linear density. The preferred units are N/tex or mN/tex, other units which are found in the industry are: gf/denier and cN/dtex.

$$\text{Specific stress} = \frac{\text{force}}{\text{linear density}}$$

Tenacity is defined as the specific stress corresponding with the maximum force on a force/extension curve. Breaking length is defined as the theoretical length of a specimen of yarn whose weight would exert a force sufficient to break the specimen. It is usually measured in kilometres. Elongation is the increase in length of the specimen from its starting length expressed in units of length. The distance that a material will extend under a given force is proportional to its original length, therefore elongation is usually quoted as strain or extension percentage. The elongation at the maximum force is the figure most often quoted. Strain expresses the elongation as a fraction of the original length. The elongation that a specimen undergoes is proportional to its initial length.

$$\text{Strain} = \frac{\text{elongation}}{\text{initial length}}$$

Extension percentage is the strain expressed as a percentage rather than a fraction.

$$\text{Extension} = \frac{\text{elongation}}{\text{initial length}} \times 100\%$$

Breaking extension is the extension percentage at the breaking point.

New Words and Expressions

yarn number 纱线支数
yarn count 纱线支数，纱线密度
nomenclature 名称，术语
hank 纱绞
spun-silk 绢丝
yarn size 纱线支数，纱线粗细
tex 特克斯
cable 粗绳索
forefinger 食指
medieval 中世纪的
untwist 退捻
curl 卷曲
kink 扭结
breaking strength 断裂强力
stress 应力
specific stress 比应力
tenacity 强度
breaking length 断裂长度
elongation 伸长
strain 应变
extension percentage 伸长率
breaking extension 断裂伸长

Notes to the Text

1. ...parallel to the center bar of the letter S：与字母 S 中间的斜杠平行。
2. unaided eyes：肉眼。
3. while the torque on the two-ply yarn will be 100 in the S direction：此处的 100 是假定的一个没有单位的数值。
4. If the same two singles are plied into a double：句中的“double”可译为“双股线”。
5. This simple analysis holds only for two-ply yarns：句中的“holds”意思是“适用，有效”。
6. yd：yard 的缩写形式。

Questions to the Text

1. Describe the twist direction of the yarn.
2. Give an example to show how to produce the balanced yarn.
3. How is the yarn number defined?
4. What is the denier system?
5. What is the tex system?
6. Describe the definition and difference about strength and elongation.

Lesson Nineteen

Yarn Properties(2)

Yarn hairiness

Yarn hairiness is usually characterized by the amount of fibers protruding out of the compact yarn body. In a spun yarn, the majority of the fiber ends are embedded in the main structure, although a few ends may protrude out as a consequence of their shorter length or higher bending and torsional rigidities. Yarn hairiness has a great influence on the sizing, weaving and knitting processes. Higher hairiness increases the cost of sizing. During the shedding operation in weaving, the hairy yarns often entangle with each other and thus hinder the creation of distinct shed which is essential for the passage of the weft or weft carrier. Hairy yarns generate fly during the knitting and obstruct the smooth functioning of the machine parts, including needles. However, yarn hairiness is a necessary evil. Too much hairiness is detrimental for the fabric appearance but a certain hairiness in the yarn is also desired so that the fabric possesses a softer feel and a warmer hand. It has been observed that the comfort parameters of textile fabrics (air permeability, moisture vapour transport and thermal properties) depend on yarn hairiness.

The basic principles of yarn hairiness measurement could be classified under the following two heads. Counting the number of fibers longer than a given reference length or longer than a set of reference lengths that are protruding from the yarn body per unit length of yarn. Measuring the cumulative length of all protruding fibers outside the yarn body per unit length of yarn. There are Shirley yarn hairiness tester, Zweigle yarn hairiness tester and Uster yarn hairiness tester. The hairiness measuring principle of the Uster tester is different from that of the Shirley and Zweigle instruments, and therefore a direct comparison cannot be made between the results obtained from these instruments.

Yarn evenness

Yarn evenness can be defined as the variation in weight per unit length of the yarn or as the variation in its thickness. There are a number of different ways of assessing it.

About visual examination yarns to be examined are wrapped onto a matt black surface in equally spaced turns so as to avoid any optical illusions of irregularity. The blackboards are then examined under good lighting conditions using uniform non-directional light. Generally the examination is subjective but the yarn can be compared with a standard if one is available; the ASTM produce a series of cotton yarn appearance standards.

Cut and weigh methods is the simplest way of measuring variation in mass per unit length of a yarn. The method consists of cutting consecutive lengths of the yarn and weighing them. For the method to succeed, however, an accurate way of cutting the yarn to exactly the same length is required.

The Uster evenness tester measures the thickness variation of a yarn by measuring capacitance. The yarn to be assessed is passed through two parallel plates of a capacitor whose value is continuously measured electronically. The presence of the yarn between the plates changes the capacitance of the system which is governed by the mass of material between the plates and its relative permittivity (dielectric constant). If the relative permittivity remains the same then the measurements are directly related to the mass of material between the plates. For the relative permittivity of a yarn to remain the same it must consist of the same type of fiber and its moisture content must be uniform throughout its length. The presence of water in varying amounts or an uneven blend of two or more fibers will alter the relative permittivity in parts of the yarn and hence appear as unevenness. The unevenness is always expressed as between successive lengths and over a total length. If the successive lengths are short the value is sometimes referred to as the short-term unevenness. The measurements made by the Uster instrument are equivalent to weighing successive 1cm lengths of the yarn.

There is a close correlation between yarn count and the capacitance tester reading; the changes of capacitance brought about by alteration of the total fibre cross-sectional area between the plates enables the automatic indication of the mean and mean deviation (*U*%) or coefficient of variation (*CV*%).

Imperfections

In addition to measuring the overall variability of yarn thickness the Uster tester also counts the larger short-term deviations from the mean thickness. These are known as imperfections and they comprise thin places, thick places and neps. The sensitivity of the eye to thick and thin places in a yarn is such that around a 30% change from the mean thickness is needed for a thick or thin place to be visible. In the instrument, therefore, only thick and thin places above these levels are counted. Neps are considered to be those thick places that are shorter than 4mm whereas areas counted as thick places are the ones that are longer than 4mm. The total volume of the nep is considered in the assessment but for the purposes of counting they are all assumed to have the same length of 1mm so that any variation in size is registered as a variation in thickness. Neps are counted at sensitivities of +140%, +200%, +280% and +400% above the mean thickness.

New Words and Expressions

protrude out 突出

torsional rigidities 扭转刚度

hinder 阻碍，妨碍，阻挡；

detrimental 有害的

air permeability 透气性
moisture vapour transport 透湿性
matt 不光亮的；无光泽的；亚光的
irregularity 不规则
consecutive 连续不断的
capacitance 电容
permittivity 介电常数
dielectric constant 介电常数
successive 连续的；接连的；相继的
the short-term unevenness 短片段不匀
thin places 细节
thick places 粗节
neps 棉结

Notes to the Text

1. Counting the number of fibers longer than a given reference length or longer than a set of reference lengths that are protruding from the yarn body per unit length of yarn：句中的“reference length”可译为“标准长度”。句中的“a set of reference lengths”可译为“一系列标准长度”。句中的“per unit length of yarn”可译为“单位长度的纱线”。
2. Yarn evenness can be defined as the variation in weight per unit length of the yarn or as the variation in its thickness：句中的“the variation in its thickness”可译为“直径的差异”。
3. About visual examination yarns to be examined are wrapped onto a matt black surface in equally spaced turns so as to avoid any optical illusions of irregularity：句中的“in equally spaced turns”可译为“均匀地，等距地”。句中的“to avoid any optical illusions of irregularity”可译为“避免光学错觉对不匀的影响”。

Questions to the Text

1. What is hairiness?
2. Please describe the disadvantage and advantage of hairiness.
3. What is the basic principle of yarn hairiness measurement?
4. How many methods can be used to test yarn evenness? And please introduce them.
5. How to define the thin places, thick places and neps?

Lesson Twenty

Textured Yarn *

语言微课堂

In the early 1950s, nylon and the polyester fibers consolidated their position as commercially important textile fibers. The continuous filament yarns of immediate post war years had characteristics different from natural or regenerated fiber yarns then in use. Fabrics made from the synthetic filament yarns had attractive and unusual properties, notably superior wear and easy care characteristics; they could be washed easily; and drip dried. But they also possessed less desirable features which restricted their textile application. They were smooth and slippery to the touch, and felt clammy when worn next to skin. They did not possess the warmth and comfort of fabrics made from natural fiber yarns.

The superiority of traditional yarns in this respect was due, in large measure, to the fact that they spun from short staple fibers. Yarns made in this way were fuller and fluffier than those derived, like the synthetic fiber yarns, from continuous filaments. Spun yarns were bulky and light, with innumerable pockets of air entrapped between the short fibers. This air was primarily responsible for the heat insulation characteristics of spun yarns, and the fabrics made from them. Also, perspiration could escape through the interstices of the yarn.

It was a natural step for synthetic fiber manufacturers to produce synthetic staple fibers by cutting or breaking the continuous filaments. Synthetic staple fiber could be spun into bulky yarns by techniques used for cotton, wool and other natural staple fibers, providing yarns with greatly improved handle and warmth, which yet retained the unique characteristics associated with synthetic fibers.

Later, techniques were developed for the modification of continuous filament fibers in such a way as to provide bulkier and more comfortable yarns without the need for cutting the filaments into staple fibers. Instead, the filaments were bent, crimped, curled or looped in various ways to create bulkier yarns possessing many of the characteristics associated with spun yarns. Processes of this type become known as texturing.

Texturing gives yarns a soft and wooly feel, and increases the warmth and comfort of fabrics. The loops and crimps entrap a multitude of small pockets of air. If air is still, it is an excellent insulator, and the fibers in a textured yarn reduce air movement to a minimum. The fibers themselves conduct heat more readily than static air, and the warmth of a fabric is related directly to the amount of entrapped air. The bulkier and softer the yarn, the warmer it will be. The filaments of synthetic fiber play only a minor role in controlling the warmth of the fabric; the static air provides insulation, the filaments preventing air movement and holding the fabric together.

Warmth is only one of the desirable characteristics of a textile material. Yarns and fabrics must also be flexible, which means that the yarn should be of fairly low linear density and constructed from fine components. Ordinary staple yarns are satisfactory in this respect, but their mechanical flexibility is restricted by their structure. Flexibility depends on the extensibility of the outermost fibers of the yarn. A straight fiber is much less extensible than a coiled, looped or crimped fiber; the extensibility of such fibers is controlled by a combination of bending stiffness determined by the geometrical shape of the fiber in the yarns.

Bending stiffness is the function of the linear density of the fiber; a fine fiber is more flexible than a coarse one. A yarn made of many looped, coiled or crimped fine fibers, loosely assembled so as not to restrict one another, would thus be flexible and very extensible. The extent of these properties will depend upon the geometric configuration of the fibers making up the yarn, their relative disposition and the extent to which they restrict one another. The softness of the yarn, thus depends on the manner of texturing. Other things being equal, the softer a yarn, the greater is its covering power and the more desirable its handle.

In general, an increase in the bulkiness of a yarn results in greater warmth, softness, flexibility and extensibility. Without texturing, a continuous filament yarn with a minimum of bulk, is deficient in these characteristics. The thermosetting polymers from which most textured yarns are made are tough and strong; the yarns therefore have good abrasion resistance and durability. These characteristics are carried forward into the fabrics, which also have good pill and crease resistance, good dimensional stability and excellent washing and drying properties. Synthetic fibers, as a rule, do not absorb moisture as well as nature fibers but the entrapped air holds moisture and bulked yarns are reasonably comfortable in this respect. They acquire static electric charges more readily, but special finishes may be used to minimize this tendency.

Textured yarns are wound onto large packages, providing long knot free lengths of a uniform product which lends itself admirably to knitting. The increase use of textured polymer filaments has proceeded side by side with a rapid increase in machine knitting, and each development has helped the other. Textured yarns and machine knitted fabrics, for example, are both capable of being produced at high speed. The knitting process provides fabric of high flexibility and extensibility; it is essentially a loop through loop process, and the yarn itself is flexible and extensible. In some ways, yarn and fabric complement one another; the high cover factor of the yarn, for example, tends to offset the poorer cover factor of the fabric.

Stretch yarns have high extensibility and good recovery, but only moderate bulk in comparison with bulked yarns. They are used mainly in stretch fabrics which are made up into many domestic goods and garments. They are produced mainly by false twist type machines.

Bulked yarns have moderate stretch, being used in applications where bulk is more important than extensibility, e. g. in carpets, upholstery, warm garments and hosiery. They are produced by stuffer-box, air-jet, and various other types of crimp producing devices.

New Words and Expressions

full 丰满，充满
handle 手感
linear density 线密度
extensibility 伸展性
bending stiffness 弯曲刚度
thermosetting 热定形
knot free 无结
cover factor 覆盖系数
offset 抵消
stretch yarn 弹性丝，弹性纱
false twist 假捻
bulked yarn 膨体纱
carpet 地毯，毛毯
hosiery 针织品，袜类
stuffer-box 填塞箱
air jet 气流

Notes to the Text

1. The continuous filament yarns of immediate post war years had characteristics different from natural or regenerated fiber yarns then in use：句中“immediate post war years”意思是“战后的年代”，“regenerated fiber”指“rayon，viscose fibers”。
2. in large measure：在很大程度上。
3. Yarns made in this way were fuller and fluffier than those derived，like the synthetic fiber yarns，from continuous filaments：句中“like the synthetic fiber yarns”为插入语。
4. innumerable pockets of air：无数的空隙。
5. wooly：羊毛般的，另一种拼法为“woolly”。
6. a multitude of：大量。
7. fine component：结构微细的原料。
8. bending stiffness is the function of the linear density of the fiber：纤维弯曲刚度是线密度的函数。
9. The extent of these properties will depend upon the geometric configuration of the fibers making up the yarn：句子中“The extent of these properties”可译为“这些性能的高低”。
10. do not absorb moisture as well as nature fibers：不像天然纤维那样容易吸收水分。
11. The increase use of textured polymer filaments has proceeded side by side with a rapid increase in machine knitting：变形聚合物长丝的应用与机械针织业同时增长。

Questions to the Text

1. What are the advantages and disadvantages of the synthetic filament yarns?
2. Why spun yarns are widely welcomed?
3. Why do some filament yarns need to be textured?
4. Why larger package is good for high production rate?

Lesson Twenty-One

Blend Yarns and Fabrics*

Blend and combination yarns and fabrics are those in which two or more generic types of fibers are used. These different types of fibers can be combined in the following ways:

(1) Two or more different types of fibers can be blended into a single yarn.

(2) Single yarns, where each single is of a different fiber type, can be plied together to form a combination ply yarn.

(3) Single or ply yarns of one fiber type can be used with single or ply yarns of another fiber type and woven or knitted into a fabric.

As many different fiber types as desired can be used, but each single yarn would be of the same fiber content without blending.

The accepted definition of a blend as cited by ASTM is "a single yarn spun from a blend or a mixture of different fiber species". According to this accepted definition, only the first type of blend or combination cited above qualifies as a blend yarn. Further, only a fabric made from such yarns qualifies as a blend fabric. Unfortunately, many consumers have come to associate the term blend with any fabric containing fibers of two or more different generic types regardless of how they have been used in making a fabric. As a rule, woven fabrics designated as blends by fabric producers employ blended yarns throughout their construction. An exception would be a woven fabric in which yarns of one fiber type are used in the warp while a different fiber type is used in the filling. Other possible combinations might occur when a blended yarn is used in one direction of the fabric and a yarn composed of only one fiber type is used in the other direction. The latter yarn may be one of the same fibers used in the blend or a totally different fiber type. Fabrics in which blended yarns are used throughout the construction are more likely to give desired performance characteristics during use and care than fabrics made with single fiber component yarns but a different fiber in different yarns.

Knit fabrics may be a true blend; they may be made of a combination of yarns of different fiber types; or they may be made of ply yarns with single components of different fiber types. Nonwoven fabrics are blends when two or more species of fiber are used in constructing the fiber mat.

Blends may be composed of differing amounts of each fiber involved. A typical fiber blend is cotton and polyester. The amount of each fiber used may be similar, giving a 50/50 blend; the polyester can exceed the cotton, as in a 65/35 polyester/cotton blend or a 80/20 polyester/cotton blend; or the cotton can exceed the polyester as in fabrics with 65 to 80% cotton. In a true blend, each yarn has the same proportion of each different fiber type involved as every other yarn.

For a majority of blends on the market, optimum percentages have been established for fibers involved. For example, textile scientists generally agree that in blends of polyester and cotton, the percentage of cotton should range between 35 and 50% if the fabric is to combine the properties of easy care, durability, and comfort. If easy care and durability are the most important, the polyester component can go as high as 80%; if comfort is the most important, the cotton content may be as high as 80%.

A blend of 55% acrylic with 45% wool, or a blend of 50 to 65% polyester and 35 to 50% wool, results in fabrics that provide comfort and can be washed instead of dry cleaned. A fabric in which as little as 12 to 15% nylon is blended with 85 to 88% wool and given proper finishing is considered to be a washable wool fabric. In addition, the nylon increases the resistance of the fabric to abrasion damage and helps prevent a high loss in strength that might occur because of the finishing effect on the wool fiber.

In the manufacture of a blend yarn, the fibers may be intimately mixed before yarn manufacturing starts or during the finisher drawing operation. For nonwovens the fibers must be blended before the fiber mat is formed. Unless a thorough blending occurs, the end product will not be uniform.

A factor of importance in building blend yarns is that the different fibers must be cut to the length required by the particular yarn making machinery to be used; if natural fibers are involved, any man-made fiber used with them must be cut to the average length of the natural fiber. If different natural fibers are used, it is important to try and use fibers of the same general length. For the most part, man-made fibers are cut to the length of cotton fibers, as most yarns are made on equipment designed for the staple length of cotton fiber.

Blends can be developed to provide consumers with special performance characteristics and/or to meet predetermined end-use requirements. They may be designed specifically for appearance; to combine appearance and performance; to emphasize easy care; to attract consumers who want to buy luxury fibers for prestige; for reducing cost; or for providing durable press and easy care properties. Some typical examples of blend fabrics and the reasons often posed for their development include polyester and cotton to provide easy care, durability and comfort; rayon and acetate to provide appearance and draping qualities; silk, vicuna, and cashmere to add prestige; linen to provide stiffness and appearance.

Blends or combination fiber fabrics are the result of considerable research, development, and testing. Instead of developing new fibers, manufacturers, during the past decades, have concentrated on mixing different fibers to produce yarns and fabrics with desirable properties. A blend that has been properly engineered exhibits properties that represent the best of each fiber involved.

Combination fabrics and yarns may produce outstanding results if the combining process creates yarns with properties that work well together. In the case of a combination yarn, two or more singles, each of different fiber types, are combined into a ply. Such yarns may give outstanding use and be easy to care for. When made into fabric, they should probably give good wear and performance. The major problem with combination fabrics occurs when the yarns in one direction are of one type of fiber while the yarns in the perpendicular direction are of a different fiber type. Such combination can produce fabrics that do not perform adequately; they may wrinkle badly in one direction or both, they may shrink in one direction, and they may require complex care procedures.

Strength is introduced into combination fiber fabrics by using yarns of high breaking load in the direction that requires extra resistance to force or by spacing strong yarns among yarns of low tenacity fibers; the strong fibers increase the resistance to breaking. Nylon yarns in the warp direction with some low tenacity fiber in the filling would produce such a fabric.

New Words and Expressions

combination yarn 混纺纱线
combination fabric 混纺织物
ply yarn 多股纱线，合股纱线
generic type 某种类型的，某种属性的
mat 毡
dry cleaned 干洗
vicuna 骆马毛
cashmere 开司米，山羊绒

Notes to the Text

1. ASTM：有两种含义，一是“American Society for Testing Materials”，美国材料试验学会；二是“American Standard of Testing Materials”，美国材料试验标准。从本文内容分析，宜译为美国材料试验学会。
2. have come to associate something with something：习惯上将……与……联系。
3. In a true blend, each yarn has the same proportion of each different fiber type involved as every other yarn：正规的混纺纱线中各种纤维含量的比例是相同的。
4. if the fabric is to combine the properties of easy care, durability, and comfort：如果织物对容易保养、耐用和舒适性的要求是等同的。
5. washable wool fabric：“washable” 特指“可水洗的”。
6. by spacing strong yarns：“spacing” 指“间隔排列”。

Questions to the Text

1. How many ways can be employed to produce blended fabrics?
2. In choosing the percentages of different fibers in a blended yarn, what are the major considerations?
3. When making a blended yarn, in which processing step different fibers may be mixed?
4. In which case problems may occur as far as combination fabric is concerned?

语言微课堂

Lesson Twenty-Two

Woolen and Worsted

Garments made from wool may be either woolens or worsted. The difference between these two classes of goods is not usually understood by people who are not intimately connected with the trade. Quite often a certain article of clothing which is labeled all wool may be incorrectly referred to by the layman as woolen, whereas in reality the particular garment should be more correctly termed worsted. It is hoped that the following explanations will help to clear up any confusion that may exist in the reader' s mind.

Woven fabrics are made from yarns, which in turn are composed of textile fibers twisted together. The basic fiber used for both woolen and worsted goods is wool, but in worsted yarns the fibers have been laid parallel to each other during manufacture giving the yarn and the ultimate fabric a neat, smooth appearance, whereas yarns in which the fibers are crossed in all directions and are not parallel, and therefore have a rough, whiskery appearance, are woolen. Cloth is named after the yarns that compose it. The woolen cloths appear full-handling and bulky, but the worsted fabrics have a clear, smooth, regular surface with the individuality of each thread apparent in the pattern of the weave.

专业微课堂：
毛呢面料的特点

Woolen is a clear enough term to the layman and signifies something to do with wool, but it is doubtful whether many people, even some workers in the industry, could say much about the derivation of the term worsted.

Worsted is a slight corruption of Worstead, the name of a village in Norfolk. It was here that the expert cloth workers who entered England in the early fourteen century as refugees from religious and political tyranny on the Continent, particularly Flanders, first settled and introduced new and original methods for the production of superior and finer cloth than had hitherto been made in Britain. Formerly a flourishing town, Worstead is now an agricultural village without an industrial spindle or loom.

Only virgin wool is used in worsted manufacture. In order to make a yarn with a clear outline the worsted trade uses wool fibers which have been combed. This process straightens the longer fibers and removes the shorter ones. All worsted yarns, except a small percentage used for making carpets and some hand knitting wools, are made from combed wool; in the manufacture of worsted-spun carpet yarn the combing process is sometimes omitted for economy and also to leave in the yarn the short fibers which contribute to the weight of the finished carpet. This system may be referred to as semi-worsted. All the machines used in worsted yarn manufacture help to maintain the smoothness of the thread.

The woolen spinner may use virgin wool also, but on the other hand his yarns may not contain any new wool at all, for in addition to pure wool he has at his disposal a wide variety of other fibrous materials including:

(1) Shoddy fibrous material produced by tearing up old and new knitted garments or loosely-woven fabrics in rag form.

(2) Mungo-produced in a similar way to shoddy but the rags used comprise new or old hard-woven or milled cloth or felt.

(3) Noils—these are pure wool of short fiber length removed from the longer wool fibers during the combing process in worsted yarn manufacture.

(4) Cotton, rayon and other man-made fibers in the staple form.

(5) Wastes made at various stages of textile processing.

Wools of short fiber length can be used in the woolen industry, indeed it is said that the low woolen trade can deal with any fiber, however short, so long as it has two ends! Short wools which lack sufficient length for combing are often termed clothing wools and are used for woolens, whereas wools which are classed as suitable in length for the combing process and used in the worsted industry are termed combing wools.

There are many more worsted than woolen processes, and it is probably true to say that in yarn manufacture it may require almost as many weeks to pass from raw material to yarn in worsted processing as it requires hours in woolen manufacture.

To produce a woolen yarn the components of the blend are first mixed together in blending machinery. The resultant blend is then processed into slubbings on a woolen carding machine, and finally the slubbings are spun into yarn on a woolen mule or a woolen spinning frame. In worsted yarn manufacturing, however, after the wool has been scoured and dried it is usually carded. It must then be combed and gilled, followed by a series drawing operations before it is finally spun. At every stage attempts are made to lay the fibers parallel to each other in order finally to produce a smooth worsted yarn. There may be as many as twenty machines in the sequence of operations for worsted yarn as against only three or four for a woolen yarn.

After weaving there are many more possibilities in the range of finishes which can be applied to woolens than to worsted. In fact it has been said that a worsted cloth is made in the loom whereas a woolen cloth is made in the finishing. Usually there is little difference in general appearance between a worsted cloth taken from the loom and the same fabric ready for dispatch to the customer, but seldom can the same be said of a woolen fabric. Indeed a woolen blanket as it comes from the loom looks bare and thready, whereas on sale in the shop it is a fibrous, soft-handling article, and all this is due to the finishing processes.

New Words and Expressions

whiskery 须状的

virgin wool 新羊毛

full-handling 丰满手感

semi-worsted 半精梳

shoddy 软再生毛
mungo 硬再生毛
milled cloth 缩绒织物，缩呢织物
noil 落毛，落纤维
clothing wool 粗梳用毛
combing wool 精梳用毛
slubbing 头道粗纱
mule 走锭纺纱机
gill 针梳
thready 能明显看出纱线的

Notes to the Text

1. Quite often a certain article of clothing：“article”指“物品”，“a certain article of clothing”可以译为“一件服装”。
2. with the individuality of each thread apparent in the pattern of the weave：纹路清晰。
3. Woolen is a clear enough term to the layman and signifies something to do with wool：“Woolen”这个词对于外行是个再清楚不过的词了，它表明了与羊毛有某些关系。
4. Worsted is a slight corruption of Worstead：句中的“corruption”表示“讹误”，句子的意思是“Worsted 是对 Worstead 的有小偏差的误称”。
5. Norfolk：诺福克，英国郡名。
6. Continent：大写时特指欧洲大陆。
7. Flanders：佛兰德地区的人，佛兰德地区指目前比利时和法国的部分地区。
8. superior and finer cloth than had hitherto been made in Britain：这些织物比英国在此之前生产的织物更高档，更细致。
9. however short, so long as it has two ends：无论如何短，总会有两个头端。
10. mule：其原意为“骡子”。由于走锭纺纱机的结构是多种机械机构综合，“杂交”而来，因此使用了该词来表示。

Questions to the Text

1. What does the worsted fabric look like? And the woolen fabric?
2. What materials are usually employed for the manufacture of the worsted goods? And the woolen goods?
3. Where is the word “worsted” from?
4. Describe the difference between the processes in making worsted and woolen fabrics.
5. Explain the sentence：“a woolen cloth is made in the finishing”.
6. Write a paper with no less than 500 words to tell the difference between worsted and woolen goods.

Lesson Twenty-Three

Raw Wool Treatment*

语言微课堂

Raw wool is contaminated with impurities which are usually allowed to remain in the wool until after it is sold in the country of yarn or cloth manufacture. The percentage of these impurities varies very considerably in different classes of wool; for example, a merino may contain 45 to 55% impurity, whereas some English wools only show a loss of 20 to 25% after the cleansing processes.

The fibers in a fleece of wool vary in length, diameter and general condition from one part of the sheep to another. In order to get the best out of a lot of wool, it must be sorted. This is a highly skilled manual process whereby each fleece is divided up into various qualities, and it involves sight and touch. Both these senses have to be highly trained to sort wool to the satisfaction of the passer, who knows exactly what is required from each lot. The work is conducted in a good light, preferably facing north, by men who have served a proper apprenticeship to this highly skilled job.

The sorter divides the fleece into various matchings, the real difference between which is the average fiber fineness or diameter. He looks for length, waviness and character of tip, and feels the wool for handle, density, strength and soundness. In addition to sorting the sound wool into various categories, he rejects material heavily contaminated with impurity. The wool sorters' matchings are given special names or quality numbers such as 70s, 64s, 58s etc., the higher the number the better the quality of the wool.

专业微课堂：
羊毛加工过程

Wool is usually scoured before being processed into yarn so as to remove grease, sweat and mineral impurity. Scouring is sometimes omitted for certain lower grade qualities intended for the carpet or blanket trade and these are carded in the greasy state. The grease left in may assist as a fiber lubricant in processing but the dirt which is also present can be objectionable. In any case the grease must be removed later, so the present-day tendency is to scour such greasy wools before the blending and carding processes, and then apply a carding oil in blending for the purpose of fiber lubrication, since this results in cleaner machinery and working conditions, as well as the elimination of possible faults in dyeing and finishing.

Raw wool is scoured in large machines, specially constructed for mass production. The cleansing of wool by the emulsion system is usually performed in sets of scouring bowls; this may be preceded by a shaking machine and/or a water steeping process to remove, as more as possible, the heavy sand and water-soluble impurities before the scouring process.

The washing set consists of a series of tanks or bowls, usually three to five in number according to the amount of impurity to be removed, separated from each other by pairs of revolving squeeze rollers; the whole apparatus makes an automatic unit for effective wool cleansing. Heated water, soap and alkali (or other detergents) are the scouring agents used, assisted by agitation from the mechanical arrangement which propels the wool through the bowls.

The wool is fed automatically into the first bowl of the series from a hopper at a rate depending on the size of the bowl and the type of wool. When it falls into the scouring liquor it is first thoroughly wetted out and is then moved slowly forward through the bowl by a set of forks, helped by a flow of water in the same direction.

Heavily-weighted squeeze rollers at the delivery end of each bowl take out the dirty water which passes into a side setting tank, where a partial separation of grease and dirt takes place, the liquor being pumped back into the main bowl. The wool emerging from the squeeze rollers passes into the next bowl for further treatment. The conditions maintained in a wool scouring set vary from firm to firm and often within a firm.

Despite the use of the heavily-weighted squeeze rollers at the final delivery of the wash bowls, the wool leaving a washing set contains about 50% of moisture which must be removed before further processing. This can be effected either ①physically by evaporation or ②mechanically by squeezing or by centrifugal force in a hydro-extractor. The later method does not reduce the moisture content sufficiently for efficient subsequent processing, and also, being an intermittent process, it is time-consuming and costly in labor. Therefore, hot air drying is usually practiced after wool scouring.

In the more primitive days of woolen manufacture when wool was washed in rivers and streams, it was presumably dried by hand-wringing or by the action of the wind and sun in the open air. Drying by hot air is popular today and is based on the fact that moisture can be transferred from the wet wool if brought in contact with heated air. Anyway, hot air in the presence of wet wool is not sufficient. Before it becomes saturated with moisture the air must be removed, or condensation of water may occur. Adequate pressure is necessary to force the air to do more than simply impinge on the surface of the wool staples under treatment; it must be sufficient to ensure that the interior of the staples is dried.

Vegetable matter, such as burrs, seeds and straw can be removed from wool either:

①Mechanically—just before and/or during carding.

②Chemically—immediately after wool scouring or during cloth finishing.

The stage chosen depends on the type and quantity, the method of processing and the resultant fabric. The worsted industry usually prefers mechanical burr removal, whilst in woolen manufacture both mechanical and chemical means are in use.

The chemical methods depend on the fact that vegetable matter is cellulosic and can be converted into a black brittle hydrocellulose by treatment with certain mineral acids or some of their salts. The process is known as carbonizing. After a normal emulsion scour, loose wool which is to be carbonized is usually passed through two tanks, lead-lined or of a special stainless steel, similar to scouring bowls, but containing cold dilute sulfuric acid; the wool is then squeezed and passed through a continuous drying machine at a high temperature to concentrate the acid and promote cellulose attack. Two crushing and shaking processes are used to pulverize and remove the brittle hydrocellulose from the wool, which is then neutralized by a bath of soda ash solution and double-rinsed in two scouring bowls.

New Words and Expressions

passer 检查员
apprenticeship 学徒年限，学徒身份
matching 拼毛，将同级套毛并在一起
soundness 羊毛的强力程度
carding oil 梳毛油
emulsion system 乳化洗毛系统
scouring bowl 洗毛槽
setting tank 沉淀桶，澄清桶
hydro-extractor 离心脱水机
burr 草籽，草刺
hydrocellulose 水解纤维素
pulverize 粉碎

Notes to the Text

1. ...and it involves sight and touch：其中包括眼观和手摸。
2. waviness and character of tip：卷曲状态以及毛尖的特点。
3. mass production：大批量生产。
4. this may be preceded by a shaking machine：在此之前可以设置一台振荡设备。
5. water steeping process：浸水工艺。
6. The conditions maintained in a wool scouring set vary from firm to firm and often within a firm：洗毛的工艺条件在不同的工厂是不相同的，即使在同一工厂也经常使用不同的工艺条件。
7. or condensation of water may occur：否则会产生凝结水。
8. lead-lined：用铅板做内衬。
9. soda ash solution：纯碱溶液，苏打灰溶液。

Questions to the Text

1. How is the wool sorted?
2. Why scouring of the wool is necessary?
3. Describe the wool washing process.
4. How to remove the remained moisture in the wool after scouring?
5. What is carbonizing?

Lesson Twenty-Four

Woolen Carding*

语言微课堂

A carding process is necessary in the production of most staple yarns, e. g. worsted, woolen, cotton, jute, flax and silk noils; but the machines used for carding the various fibers differ in construction. The reasons are:

(1) In the woolen industry carding follows the blending operations, and after it there is only one further process in yarn manufacture, that is, spinning.

(2) Materials used in woolen industry are special. The woolen trade can use shorter wools. The length of fiber partly determines whether a particular wool is suitable for processing on the worsted system or whether it is more suitable for the woolen trade. In addition to pure wool, re-manufactured materials are widely used in the woolen industry. In general these comprise shoddy, mungo, extract, noils, flocks and various textile processing wastes. Extract is derived from cloths comprising cotton-and-wool; the cotton is removed (extracted) by carbonizing and the wool which remains is pulled, as for shoddy and mungo.

Therefore extreme care and vigilance must be exercised in carding, because the spinning process can do little if anything at all to rectify faults made during carding.

The objects of woolen carding may be stated briefly as:

(1) To disentangle the fibers from the different locks of wool or clusters of other fibrous material and, at the same time, continue the work that has already been started in the blending room, by blending together these disentangled fibers so as to produce a mixture of fibers as uniform as possible.

(2) When applicable, to mix together different colors and/or qualities of fibers into a state as homogeneous as possible.

(3) To remove from the fibers the maximum amount of impurity.

(4) To produce continuously a web of fibers, uniform in thickness both across the web and throughout its length. To divide this web into a number of narrow ribbons of fibers, say one hundred, each as uniform as possible in thickness and width, and to consolidate these ribbons of fibers into cylindrical form called slubbings ready for spinning.

In short, woolen carding converts locks or clusters of fibers into a continuous form suitable for spinning into a yarn.

The machine used to accomplish these objectives consists essentially of a number of large cylinders or swifts and smaller rollers equal in width, usually 150 - 180cm but of various diameters, all covered with a material called card clothing, which consists of a firm foundation on either leather or several layers of fabric-with-rubber or similar material, in which short wires of hardened and tempered mild steel are densely set, to extent of about 15–110 points or more per square centimeter according to their position on the machine.

The carding machine (or card as it is usually called) consists of a number of units composed of a series of rollers comprising a swift with workers and strippers, a doffer and a fancy; the number of such units in a complete machine depends mainly on the type of material being processed, the kind of card clothing used on the card, and the desired result as regards yarn appearance, thickness and quality. The card is always divided into two or more sections, called scribbler and carder in a two-section machine, and scribbler, intermediate and carder in a three-section card. The object of this division into several sections is to provide better mixing of the fibers by means of the intermediate feeds which connect the sections.

At the input end of a carding machine the rollers run slowly and have comparatively few but strong wire teeth. As the wool progresses through the machine the pinning becomes denser and the wires finer; also the various rollers can be run faster. The relative speeds, direction of revolution and inclination of the wires, together with the distance apart (setting) of the adjacent rollers, determine the degree of opening and mixing given to the fibers. Other contributory factors are the sharpness of the wire teeth, the suitability of the wire diameter and the density of pinning in the card clothing for the particular type of wool, the efficiency of the precious wool scouring and drying operations, and the atmosphere conditions prevailing in the carding shed.

Even everything possible is done to avoid or minimize the breakage of fibers and to maintain fiber length throughout processing, some fiber breakage is inevitable. The reasons for this are beyond the scope of this text.

The type of card clothing used in the woolen industry varies considerably according to the kind of blend being processed, the composition of the carding set, and also the personal ideas on the man-in-charge, the carding engineer.

Card clothing may be either fillet or sheet. The former is in endless strip form varying from 2.5–7.6cm in width, usually of vulcanized cloth to hold the teeth firmly. It is wound tightly round the rollers to give an unbroken surface of card teeth. The second is not now very common, consists of leather sheets of a standard width—usually a 14.3cm wide sheet with 12.7cm covered with teeth—and of a length equal to the width of the card. These sheets are nailed on to the roller side by side across the card. Being easy to fix they are often used on the swifts, doffers, workers and fancies.

The wire teeth in carding are usually in the form of staples bent at the middle, each staple having two points. The staples pass through the foundation of the card clothing and are held firmly against the metal roller by the tightness of the cloth foundation.

In the carding machine the pairs of adjacent rollers do not actually touch each other but are set by the carding engineer according to the particular type, quality and condition of wool, by means of special thin metal gauges so that they are a known distance apart. Only with the fancy rollers should actual contact be made; the teeth of the fancy rollers mesh into the teeth of the swifts.

In a carding machine where the rollers are operating with their teeth point-to-point, an opening action occurs or, in technical language, working is done. There is in effect a contest between the two rollers for possession of the wool, during which any material held momentarily by either roller is combed through by the teeth of the other roller, resulting in clusters of fibers being disentangled, pulled out or straightened.

New Words and Expressions

lock 毛撮
cluster 一簇羊毛
homogeneous 均匀的，同质的
swift 大锡林，大滚筒
worker 工作辊，梳毛辊
stripper 剥毛辊，剥棉辊
doffer 道夫
fancy 花式的，风轮
scribbler 预梳机
carder 梳理机
intermediate 二道梳毛机
fillet 钢丝针布
vulcanize 使硬化，使硫化

Notes to the Text

1. tempered mild steel：经过回火的低碳钢。
2. to extent of about 15−110 points or more：句中“points”指“梳针”。
3. the pinning becomes more dense and the wires finer：句中“pinning”指梳针的设置。
4. the atmosphere conditions prevailing in the carding shed：梳毛车间的空气条件。
5. to give an unbroken surface of card teeth：句中“unbroken”指梳针要连续。

Questions to the Text

1. What are the purposes of the woolen carding process?
2. What are the constructions of the carding machine?
3. How does the woolen carding machine work?

Lesson Twenty-Five

Yarn Winding

语言微课堂

The yarn must be repackaged to meet the particular demands of the fabric forming system. The purposes for winding yarn are ① to produce a package which is suitable for further processing, and ② to inspect and clear (remove thick and thin spots) the yarn. A winder is divided into three principal zones: ① the unwinding zone, ② the tension and clearing zone, and ③ the winding zone.

To rewind the yarn on a new package, it must first be removed from the old package. This is accomplished in the unwinding zone. Over end withdrawal method is the simplest and most common method of yarn withdrawal. The package need not be rotated as the yarn is pulled over the end of the package. Two factors must be taken into account when this method of withdrawal is used. The first is ballooning. As the yarn is unwound from the package at high speed, centrifugal forces cause it to follow a curved path. As the yarn rotates, it gives the illusion of a balloon above the package. This ballooning leads to uneven tensions which may or may not alter some of the particular properties of the yarn. The second factor is that for each time one complete wrap of yarn is removed from the supply package, the twist in that length changes by one turn. For most yarns this change is insignificant and may be ignored. However, some fabrics are constructed using flat yarns of metal, polymers or rubber. In these cases, the yarn must remain flat and even one turn of twist is unacceptable.

The next zone is the tension and clearing zone. It is in this zone that the yarn receives the proper tension to provide an acceptable package density and build for further processing. This zone consists of a tension device, a device to detect thick spots, or slubs, in the yarn and a stop motion which causes the winding to stop in the case of yarn break or the depletion of a supply package. The purpose of the tension device is to allow the maintenance of proper tension in the yarn in order to achieve a uniform package density. The tension device also serves as a detector for excessively weak spots in the yarn which break under the added tension induced by the tension device. Tension devices fall into three categories: ① capstan tensioner, ② additive tensioner, and ③ combined tensioner. The capstan tensioner is expressed in Fig. 25. 1. The capstan tensioner depends upon the coefficient of friction between the yarn and the post, denoted by μ, and the angle of wrap, denoted by θ. If the incoming tension is denoted by T_{in} and the outgoing tension by T_{out} then, for the capstan tensioner, the relationship between the incoming and outgoing tensions is given by:

$$T_{out} = T_{in} e^{\mu\theta}$$

where e is approximately 2.718, the base of the natural logarithms. The following observations may be made about the capstan tensioner:

(1) Since μ, θ and e are constants, the outgoing tension is merely a constant multiple of the incoming tension, hence the name multiplicative tensioner.

(2) If the incoming tension is zero so is the outgoing tension.

(3) To vary the tension, at least one of the following must be done: the coefficient of friction, the angle of wrap, the number of posts and the incoming tension.

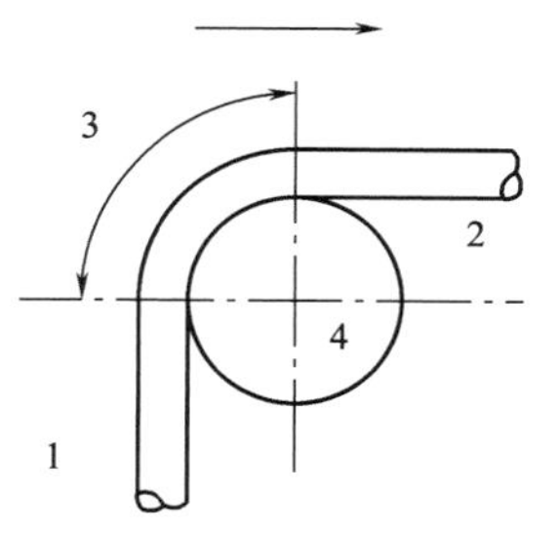

Fig.25.1 Capstan tensioner

1—the in tension 2—the out tension 3—wraping angle (θ) 4—post

专业微课堂：
络筒机

The additive tensioner depends upon the coefficient of friction between the weighting plates and the yarn, μ, and the force applied to the yarn by these weights, F. The relationship between incoming and outgoing tension in the additive type tension device is given by:

$$T_{out} = T_{in} + 2\mu F$$

Fig.25.2 is a diagram of the additive tensioner.

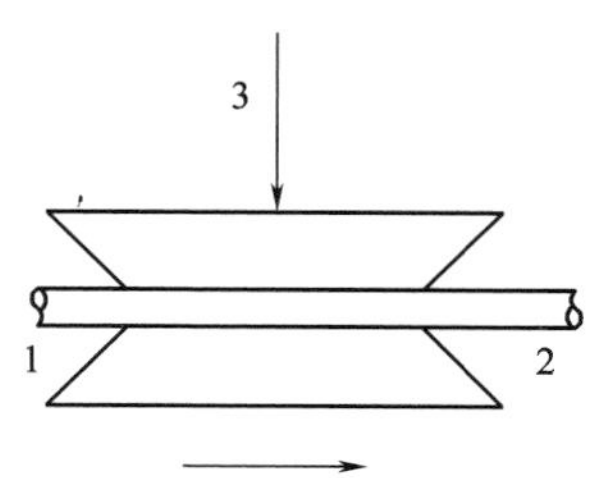

Fig.25.2 The additive tensioner

1—incoming tension 2—outgoing tension 3—weight F

The following observations may be made concerning additive tensioners:

(1) Since μ, F and 2 are all constants for a given system, the outgoing tension is simply a constant added to the incoming tension, hence the name additive.

(2) If the incoming tension is zero, there is still an outgoing tension $2\mu F$.

(3) The outgoing tension may be changed simply by changing the weight F.

The most common type of tensioning device found on winding machines is the combined tensioner. This device consists of a capstan tensioner which accepts weight discs and thus also functions as an additive tensioner.

Upon leaving the tension device, the yarn passes through a detector whose purpose is to detect thick spots. This detector may be as simple as a frame which contains an adjustable blade which can be set to allow only predetermined yarn diameters to pass through. This device is often called a snick blade. The detector, however, may contain sophisticated electronics which continuously monitor the yarn to detect thick (or thin) portions. After leaving the slub catcher, the yarn passes through a stop motion device. The purpose of the stop motion is to stop winding when the yarn breaks or runs out.

The yarn is now ready to be put on a suitable package in the winding zone. This package may be one of many types, a cone, a tube, a cheese, a dye tube or a spool, depending upon the next operation the yarn must encounter. The yarn is wound on the package by only rotating the package. This rotation may be accomplished in one of two ways:

(1) Spindle drive, where the spindle upon which the package is placed is driven directly.

(2) Friction drive, where the spindle upon which the package is placed is free to rotate and the package is driven through friction, by contacting with a driven drum.

Spindle drive winders consist of two types, constant speed winders and variable speed winders. For the constant speed spindle winder the angular velocity of the package, ω, is constant. As more yarn is wound onto the package, R, the package radius, increases. This in turn, recalling the relationship between yarn speed, driving speed and radius, causes an increase in v. Since the tension on the package is a function of the yarn velocity, a change in v causes a change in package tension and, therefore, the tension is unequal throughout the package. To overcome this drawback in spindle drive winders, a variable speed winder is used. This spindle speed is not constant but varies with package radius to ensure the stability of yarn velocity.

There is a simpler way to accomplish the same task and that is the use of a friction drive winder. In this type of winder the package is driven by a constant speed friction drum. The yarn passes between the friction drum and the package and is taken up by the package. At the point of contact of the package, drum and yarn, if no slippage occurs, all three must have the same velocity. If the yarn velocity is denoted by v_y, the drum velocity by v_d, the drum radius by R_d and the drum angular velocity by ω_d hence (Fig.25.3):

$$v_y = v_d = \omega_d R_d$$

But note that: ① the friction drum is rotating at a constant speed thus ω_d is constant, and ② the radius of the friction drum remains constant. Therefore, since R_d and ω_d are constant, v_d and v_y hence are always constant. So, for the friction drum type winder, constant yarn speed may be achieved without resorting to variable speed devices of any sort.

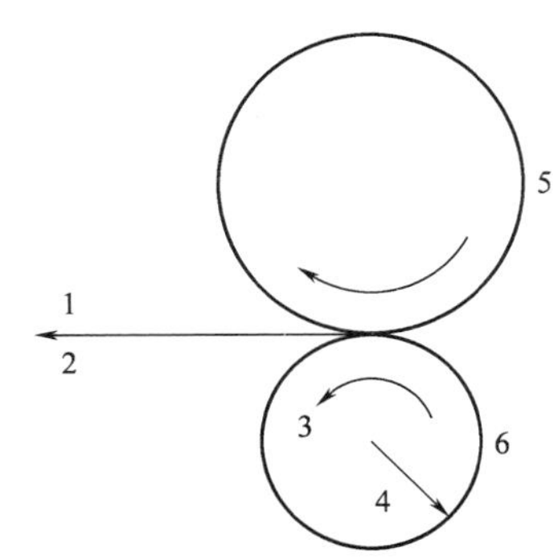

Fig.25.3 Yarn velocity

1—v_y 2—v_d 3—ω_d 4—R_d 5—package 6—drum

Not only must the yarn be wound on the package but also it must be distributed evenly along the length of the package. This is the function of the traversing mechanism. A method of traverse found only on friction drive winders is the use of a traversing groove cut into the friction drum. In this method of traverse, the yarn rides in the groove in the friction drum and is carried back and forth along the length of the package.

New Words and Expressions

winding 络纱
package build 卷装成形
cheese 扁柱形筒子
cone 圆锥形筒子
thick spot 粗节
thin spot 弱段
winder 络筒机
unwinding 退绕
rewind 重新卷绕
spool 筒子
over end withdrawal 轴向退绕
ballooning 气圈
centrifugal force 离心力
wrap 一圈，缠绕
slub 纱线粗节
depletion 耗尽
capstan tensioner 柱式张力器，倍积式张力器
additive tensioner 倍加式张力器
combined tensioner 联合式张力器
post 立柱
multiplicative tensioner 倍积式张力器
weighting plate 张力盘
weight disc 张力盘
snick blade 清纱板
slub catcher 清纱器
angular velocity 角速度
linear velocity 线速度
tangential velocity 切向速度
friction drum 摩擦辊
traversing mechanism 往复导纱机构

Notes to the Text

1. it gives the illusion of a balloon：显示出一个圈状的影像。
2. ...the twist in that length changes by one turn：在这段纱线上捻度会增加或减少一捻。
3. the base of the natural logarithms：自然对数的底。
4. Since the tension on the package is a function of the yarn velocity：句子中“function”为“函数”。

Questions to the Text

1. What are the main purposes of winding?
2. What are the basic constructions of the winder?
3. How many tension types are available?
4. Describe the tension change when using the capstan tension device.
5. How does an additive tensioner work?
6. How does the combined tensioner work?
7. Describe the work principle of the yarn detector.
8. How is the constant yarn speed achieved in the friction drum type winder?

语言微课堂

Lesson Twenty-Six

Warping

专业微课堂:
整经工艺操作流程

If the fabric forming system is weaving or warp knitting, some or all of the yarns forming the fabric are presented in sheet form. It is necessary therefore to remove the yarn from the winding package and arrange the desired number on a package called a beam. The yarns must be parallel and under uniform tension. This, is the purpose of warping.

Before thinking about winding a specified number of yarns on a beam, first consider the problem of positioning the package from which the yarn is taken in such a manner so as to facilitate the removal of yarn. Also keep in mind that the number of yarns per beam is in the hundreds or thousands and that there must, at least, be one supply package for each of these yarns. It is logical, therefore, to build a frame of some sort to hold the supply packages in a manner so as to facilitate warping. To accomplish this purpose, creels are equipped with package holders on which the supply packages are placed, tension device to help maintain uniform tension throughout the creel, guides to direct the yarn and to help keep the ends apart, antistatic devices to eliminate static charges by the rubbing of the yarn against the various surfaces, and stop motions to detect broken ends and/or empty packages.

In theory, the size of the creel (and therefore the number of packages it may hold) is unlimited. In practice, and not considering purchase price, the creel size is limited by two factors. The first of these is floor space. A creel must be housed in the building and therefore it necessarily uses some facilities of that building. Since the creel produces nothing tangible to offset the cost of housing and maintenance, it is important to consume as little of these as possible. The second factor is the yarn itself. In theoretical discussions, yarn weight, especially for short lengths of yarn, is neglected. In considering a very large creel, it is obvious that some of the supply packages must be very much further away from the point where the beam is being formed than others. Also, the yarn must be supported to keep it from dragging on the floor and tangling. Each support acts as a capstan tension device. Thus, it is important to keep the packages in a distance range where the effect of yarn weight and the effect of supports as tensioners may be neglected. Hence the size and the capacity of the creel are limited. In general, maximum creel capacity ranges from about 300 packages for very heavy yarns to 1,400 packages for thin yarns. As will be seen later, creel capacity is an important factor in warping.

Creels may be classified by the number of creel positions per end supplied. Using this classification, creels are either single or multiple package creels. To achieve higher beaming efficiency, single package creels are often used in various combinations. If the winding head, or headstock, is fixed, often non-stationary single end creels are moved in and out of position as required. These creels are called truck creels. Truck creels require that floor space be reserved for the empty creel. A more space efficient set up results if the headstock is capable of being moved. Creels used in this manner are known as duplicated creels. The advantage of a duplicated creel is the lack of need for any empty creel space in which to move an expended creel.

In one type of multiple package creel, known as a magazine creel, more than one package is provided for each end. The packages are tied head to tail so warping can continue when one package is exhausted. Another multiple package creel is known as the traveling package creel. Instead of moving creels or headstock when fresh packages are required, the packages themselves are moved into position. With a traveling package creel, the replacement of empty packages with full ones, or creeling, is done in the center while the packages in use are on the outside.

The yarn is now ready to be put on a beam. The manner in which this is done depends upon the capacity of the creel, the number of ends required in the final beam and the necessity, if any, of maintaining a pattern in the warp, e. g., for warp strips in the fabric. If the creel capacity is sufficiently high and the total number of ends required is sufficiently low or, if creel capacity is not sufficient to supply all the required ends and no distinct yarn pattern is required, then beam warping is generally used. Beam warping is simply the winding of yarns directly from the supply packages onto a beam. This beam is called a section beam since, except for the case in which all the required ends can be put on a single beam, it contains only a section of the warp required.

If, however, with insufficient creel capacity, it is necessary to build a warp beam containing the totality of ends required or if the warp yarns have to be arranged in a definite order, then drum warping is used. In drum warping, the warp is not wound directly from the creel onto the beam but rather sections of the warp are wound onto a pattern drum. In this manner the entire warp is built up in a series of sections on the pattern drum. When the total number of warp ends required in the fabric has been wound on the pattern drum, they are all removed simultaneously and wound upon a beam. This beam contains the exact number of ends required in the warp. Also, because when the ends are taken from the creel and wound on the pattern drum, exact placement in relation to each other may be made.

The final beam maintains this placement, and hence any pattern in the warp. In general, for warp knitting, the yarn for the entire fabric is not put on a single beam, but rather put up on a series smaller section beams which contain only a portion of the ends required for a full width fabric. These beams may be produced either by beam or drum warping methods. If the yarn is to be used for warp knitting, it is usually ready at this point to go to the next knitting machine. If, however, the yarn is to be used in weaving, it generally must undergo one further operation, slashing.

New Words and Expressions

package holder 筒子座，卷装握持器

antistatic device 抗静电装置

single package creel 单式筒子架
multiple package creel 复式筒子架
winding head 整经机车头
headstock 车头
non-stationary single end creel 移动单式筒子架
truck creel 转向筒子架，车式筒子架
duplicated creel 双联整经筒子架
magazine creel 复式筒子架
traveling package creel 复式移行筒子架
creeling 换筒
beam warping 轴经整经
section beam 分批整经轴，分条整经轴
drum warping 分条整经
pattern drum 分条整经大滚筒
section 经纱条带
slashing 上浆

Notes to the Text

1. A creel must be housed in the building：句子中“be housed”可译为“安装”。
2. Since the creel produces nothing tangible to offset the cost of housing and maintenance：由于筒子架不会产出任何实质性的东西来补偿安装和保养费用。
3. Creels may be classified by the number of creel positions per end supplied：筒子架可以按照为每根经纱提供的供纱位置数来分类。
4. A more space efficient set up results if the headstock is capable of being moved：句子中“results”是动词，意思是“产生”。
5. The advantage of a duplicated creel is the lack of need for any empty creel space in which to move an expended creel：使用双联筒子架的优点是：不需要为空筒子架设置空间。(由于车头可以移动，只需要一个工作筒子架和一个备用筒子架。如果车头不能移动，则需要一个工作筒子架，一个备用筒子架，还要保留一定空间，以便工作筒子架用完后移出工作区后使用。)
6. The packages are tied head to tail：筒子上的纱线首尾相连。
7. With a traveling package creel, the replacement of empty packages with full ones, or creeling, is done in the center while the packages in use are on the outside：句中“is done in the center while the packages in use are on the outside”意思是需要换筒的空筒子移动到筒子架的中部，而正使用的筒子位于筒子架的外侧。
8. The final beam maintains this placement, and hence any pattern in the warp：最后在织轴上会保持这种排列，这样任何经纱的花形(色纱排列)都会保持下来。

Questions to the Text

1. What is the purpose of warping?
2. Why the creel cannot be made too large?
3. How many types of creels are there?
4. What is beam warping?
5. In which case drum warping is recommended?
6. Describe ways to replace the empty packages in the warping machines.

Lesson Twenty-Seven

Warp Sizing

语言微课堂

In the weaving process, the warp yarns are subjected to rubbing and chafing against metal by being threaded through drop wires, heddles and reed; they are constantly being rubbed together during shedding; they are subjected to tension both constant, by the let off and take up, and intermittent, by the shedding and beat up. All of these lead to conditions which are favorable to end breakage, an occurrence which should be minimized. Furthermore, one thinks of yarns as smooth cylindrical objects when, in fact, most spun yarns are quite "hairy". During shedding these yarns move back and forth past each other. This encourages the "hairs" on one yarn to tangle with the "hairs" on its neighbors. This tangling can either lead to the tangling of the yarns themselves resulting in warp breakage, or can cause the yarns to weave as one, causing a fabric defect. Thus it is beneficial to make the outer surface of the yarn smooth. Thus, it is desirable to produce as high a quality warp as possible, one which will withstand the rigors of weaving. This is the purpose of slashing or warp sizing.

It is accomplished by:

(1) Enhancing the strength of the yarn by causing the fibers to adhere together.

(2) Making the outer surface of the yarn smooth.

(3) Lubricating or waxing the yarn to reduce friction.

In general, single spun yarns must be slashed for all the above reasons. Continuous filament yarns, if they are slashed at all, usually need adhesive to protect the filament from breaking. Ply yarns are usually slashed for lubrication and/or smoothness. In general, size ingredients can be divided into four categories:

(1) Adhesives—adhesives available include all types of starches, carboxymethyl cellulose (CMC), polyvinyl alcohol (PVA) and others.

(2) Lubricants—lubricants may be oils such as mineral and vegetable, waxes such as mineral, vegetable and animal or animal fats.

(3) Additives—additives may be included to provide features such as static elimination and mildew resistance.

(4) Solvent—the solvent generally used is water.

As can be seen, there are many possible ingredients available for a size recipe. Since slashing is a nonproductive and protective measure, it is important to carefully select the size ingredients. Some factors which must be considered are:

(1) Cost of the ingredients.

专业微课堂：
浆纱机

(2) No damage to the yarn.
(3) Compatibility with equipment.
(4) Easily removed, if necessary.
(5) Providing good fabric characteristics if not removed.
(6) No hazardous.
(7) Least amount of dusting off during weaving.
(8) Fewest number of end breaks at weaving.

Many factors influence the impact of the size upon the yarn. These factors include the size recipe and temperature, the condition of the equipment, and the amount of size picked up by the yarn. The perfect size recipe would consistently eliminate all loom warp stops and lend all desirable secondary characteristics to the finished cloth.

思政微课堂：
浆纱机的匠心之路

In general, warp sizing machines or slashers can be divided into five different sections: ① Beam creel, ② Size box, ③ Drying section, ④ Yarn separation section, ⑤ Headstock. The beam creel is merely a device or frame on which warp beams are placed in a manner convenient for unwinding. The creel can hold as few as one beam and, usually, as many as fourteen. The creel of the slasher holds all these beams and the ends on them are combined during the sizing operation. After the slasher, many warp beams are combined to form a single weaver beam. The yarn next enters the size box. The size box contains the size solution, known as size liquor. The yarn is fed into the size box by means of a guide roll. It then passes under a dip or immersion roll. This roll is capable of being moved up or down allowing the yarn to be held in the size liquor for a desired period of time. The warp sheet then passes through two rolls known as squeeze rolls. The purpose of the squeeze rolls is twofold:

(1) To squeeze out excess size.
(2) To physically drive the size into the yarn for proper penetration.

The percentage of size by weight is controlled by the yarns exposure to the size solution, governed by the speed of the machine and the immersion roll depth, the yarn structure and the pressure applied by the squeeze rolls. Fig.27.1 is a diagram showing the slasher operation.

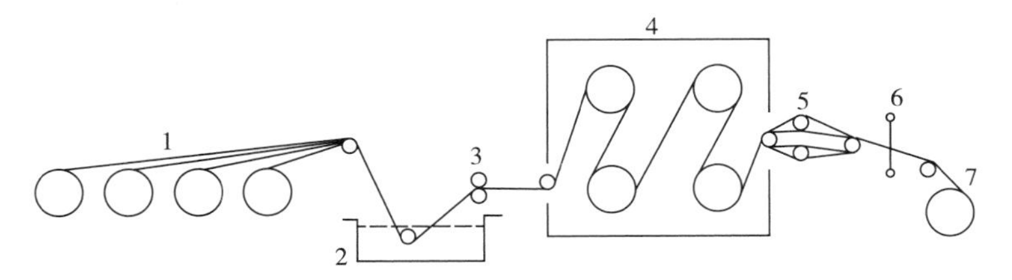

Fig.27.1 Diagram of the slasher

1—beam creel 2—size box 3—squeeze rolls 4—drying section 5—burst rods 6—zig zag comb 7—loom beam

After the yarns have been exposed to the size liquor and have picked up the required amount of size, the size solvent must be driven off, i. e. , the yarns must be dried. This drying may be done by exposing the yarns to hot air, by passing them over heated cylinders (or cans), by exposing them to infrared or by a combination of these methods. After the sized yarn has been dried it is still not ready to be put on a beam, care must be taken to separate individual ends. This task is performed by the burst rods. These are positioned on the slasher and the yarns are threaded in such a way that alternate ends are sent in alternate directions.

For a size formula to approach perfection in a given application, it should have several basic characteristics which will serve to eliminate warp breaks during weaving.

(1) The tensile strength of the yarn must be increased by the fiber-to-fiber cementing of the size film, as well as by the addition of the strength of the film itself.

(2) Adhesion is not only necessary in the fiber-to-fiber cementing, but also important to assure that the film formed by the size formula will adhere to the fiber substrate and protect the warp yarn in the weaving operation.

(3) Flexibility is particularly important in weaving. Sized warp yarns must be able to withstand repeated and extensive bending in the weaving operation without damage to the size film.

(4) Sizing of monofilament and spun yarn is a coating process. Penetration of spun yarns should be no more than sufficient to anchor the coating to the yarn, otherwise the yarn will become more like a stiff rod, losing the flexibility so necessary for good weaving.

(5) The viscosity of the size mixture directly controls the amount of penetration at a given temperature and roll pressure. The size mixture should have enough viscosity so that it will be thick enough to form the film, and at the same time lay the fibers of the yarn properly.

(6) Size formula hygroscopicity attracts moisture in the air and provides for adequate moisture regain in the warp yarn. It will undoubtedly play an important part in the lowering of humidity of the mills. Low humidity levels increase machinery longevity and stability as well as operator comfort and efficiency.

(7) Most of the troubles in any textile process can be traced to a lack of uniformity. Obviously, the weave room will run better if it is not constantly necessary to allow for variations in the sized warp. The size film must be able to withstand excessive heating caused by slowing or stopping the slasher for doffing or to cut laps. Size formula ingredients must be compatible with each other so that the addition of one ingredient will not decrease the favorable properties of another, nor precipitate it from solution.

New Words and Expressions

warp sizing 经纱上浆
drop wire 经停片
heddle 综丝
reed 钢筘
let off 送经
take up 卷取
shedding 开口
beat up 打纬
rigor 苛刻
carboxymethyl cellulose(CMC) 羧甲基纤维素
adhesive 黏合剂

polyvinyl alcohol(PVA)　聚乙烯醇
lubricant　润滑剂
additive　助剂
solvent　溶剂
size recipe　浆料配方
compatibility　适应性，相容性
dusting off　落浆
slasher　浆纱机
beam creel　经轴架
size box　浆槽
weaver beam　织轴
size liquor　浆液
guide roll　导纱辊
immersion roll　浸没辊
squeeze roll　压浆辊
heated cylinder　热风烘筒
infrared　红外线
burst rod　分绞棒
loom beam　织轴
cement　黏合
substrate　底层
hygroscopicity　吸湿性
duct　管道
longevity　寿命

Notes to the Text

1. ...or can cause the yarns to weave as one：或者使多根纱线作为一根进行交织。
2. ...if they are slashed at all, usually need adhesive to protect the filament from breaking：如果它们完全经过上浆，目的是保护长丝，不使其断裂。
3. since slashing is a nonproductive：意思是浆纱工序不直接生产产品。
4. Many factors influence the impact of the size upon the yarn："impact"指浆料与纱线结合的紧密程度。
5. ...and lend all desirable secondary characteristics to the finished cloth：为织物提供附加的优异性能。
6. The creel of the slasher holds all these beams and the ends on them are combined during the sizing operation：本句子包括两层意思，一是经轴架装载着所有的经轴；二是在上浆过程中所有的经纱汇聚在一起。两层意思用"and"连接。
7. and at the same time lay the fibers of the yarn properly："lay"的意思是"贴伏"。句子的意思是使纤维在纱线表面贴伏。
8. cut lap：指浆纱过程中割断纱线，更换织轴。
9. ...nor precipitate it from solution：不使别的成分沉淀。

Questions to the Text

1. Why warp yarns need to be sized?
2. What may be included in a size ingredient?
3. What are the major considerations in selecting the size ingredients?
4. What would be the results if the yarn contains too much size? And how about too little size?
5. What are the main constructions of slasher?
6. What are the functions of squeeze rolls?

7. How many methods can be used to dry the sized yarn?
8. Why penetration of the size into the yarn must be carefully controlled?
9. Why the viscosity of the size liquor is important?
10. How to promote the size penetration?

Lesson Twenty-Eight

Elongation and Elasticity of the Sized Yarn*

These two properties of yarn are often confused and misunderstood by even good slasher men. To define the difference: elongation is the amount of stretch the yarn will undergo before it breaks. This is generally expressed as a percentage of its length. For example 36cm of yarn, under just enough tension to keep it straight, which stretches to 37.8cm before breaking, would have 5% elongation (1.8 divided by 36). Elasticity on the other hand is the property of the yarn which allows it to return almost to its original length after the stretching tension is released.

Many of the same factors in the slashing process contribute to both elongation and elasticity. Humidity, or the amount of moisture in the yarn, greatly influences both elongation and elasticity. This also holds true for the amount of moisture or relative humidity in the weave room. Up to a point, the strength and elongation of cotton yarns increase as humidity rises. However, excessively high humidity (above 88%) does not usually lend itself to maximum efficiency at the loom. This is primarily because the elasticity and strength of size film coatings on cotton yarns tend to deteriorate when humidity exceeds 80 to 85% relative. In effect the strength of the size film starts decreasing above 85% relative humidity while the cotton fibers continue to gain strength. Thus a given yarn may have a greater break factor at higher humidity while the size film may soften and cause extra abrading or balling of the fibers (fuzz balls or lint buttons) during weaving.

While the above is true for cotton yarn, rayon sometimes acts almost the opposite for this fiber is stronger when it is driest. On rayon, the size film will become too brittle and lose much of its flexibility when dried sufficiently to give rayon fibers maximum strength. Also, overly dried rayon builds up excess static electricity and causes difficulties in slashing and in the weave room. For filament rayon and woven wire, tensile strength is not the most important factor in obtaining weaving efficiency. Stretch is another factor which contributes vastly to elasticity and elongation. Stretch occurs in the fiber, the size film and in the yarn itself. It is generally agreed that the amount of elongation and elasticity present in the sized yarn are directly related to the physical properties of the fiber or blend of fibers and the properties of the sizing materials used. Two extremes are glass fiber which has almost no elongation or elasticity and polypropylene which has much of both. The sizing material for best results should, as nearly as possible, have the same percentage elongation and elasticity as the yarn. That is the primary reason for so many new types of sizing materials on the market today.

New sizing materials are constantly being developed to match the characteristics of newly developed fibers and filaments. It would be unwise to try to size glass fiber with a very elastic size material because the fibers would rupture before the size film was very effective. Since it is practically impossible to match the elongation properties of a given warp yarn with a size film that has identical elongation properties, it is generally agreed that the more elasticity inherent in the size film, the more protection will be afforded. This stands to reason since greater film elasticity permits a greater absorption of stress at the fiber-to-fiber bonding point. Film which has little or no elasticity has a tendency to break or stretch to the point of break while highly elastic film-like a rubber band—extends under high tension and retracts as tension lessens. Highly elastic film thus provides repeated protection under repeated stress.

Creep is another factor affecting both the elongation and the elasticity of yarns. Most of yarns exhibit the property of creep when under tension or when relaxed after tension. A rayon yarn held under tension will yield up an extra amount of stretch during the tension period or, if it is stretched a certain distance, will relax so that less tension will be required to hold it there. That characteristic is known as creep and is considered to be the cause of set marks, and many other imperfections in the face of the cloth, after weaving. An example of other imperfections caused by creep would be crosswise irregularities which show up in the cloth when the loom is stopped for any length of time. When fibers have been under tension for some time, and tension is then released, the creep effect prevents the fibers from returning immediately to their original length. They remain slightly stretched. The complete return occurs over a period of time as fibers creep back to their original dimensions. The extent to which a yarn will creep often determines its value particularly in woven dress goods which must hold their shape. While the term creep has to do with stretch of the yarn and fibers, we normally think of stretch in terms of tension at the slasher during the sizing process.

A warp that has been stretched too much on the slasher will not have the necessary elasticity for weaving. If the yarns do not have sufficient residual elongation after slashing, the beat up of the reed will stretch them beyond their limit. The result will be a break and warp stop. Sizing normally decreases the elongation of the yarns, but a properly sized yarn would have as small a percentage decrease compared with the unsized yarn as possible. An unusually large decrease in elongation is often due to mechanical misadjustments in the slashing operation. Between 1% and 2% stretch on the slasher is usually normal for cotton operations. A 3% stretch (or more) may be the norm with other fibers or blends of fibers. To obtain certain types of constructions, more stretch than normal will often give better face on the cloth after weaving because of the twill lines. This is particularly true in drills, twills and gabardines where a slack end, not stretched tight enough in the creel to make it uniform with the rest of the ends, would show up like a railroad track in the finished cloth of any fabric which is supposed to have a twill line or definite characteristic pattern.

New Words and Expressions

balling 起球
lint 棉绒
creep 蠕变，塑性变形
set mark 开车痕
norm 标准
drill 斜纹布，卡其

gabardine　华达呢

Notes to the Text

1. Many of the same factors in the slashing process contribute to both elongation and elasticity：浆纱中的许多同一参数会同时影响伸长和弹性。
2. This also holds true for...：这一点对于……也是适用的。
3. Up to a point, the strength and elongation of cotton yarns increase as humidity rises：句中"Up to a point"意思是"在一定的范围内"。
4. Thus a given yarn may have a greater break factor at higher humidity while the size film may soften and cause extra abrading or balling of the fibers：这样，一种纱线由于含湿量高，引起浆膜变软，纤维受到过度摩擦或起球，会成为一个较重要的断头因素。
5. ...to give rayon fibers maximum strength：为了使黏胶纤维得到最大强度。
6. This stands to reason：这样说是有道理的。
7. if it is stretched a certain distance, will relax so that less tension will be required to hold it there：意思是如果将纱线拉伸，由于塑性变形，纱线会变松弛。

Questions to the Text

1. What is elongation?
2. What is elasticity?
3. How does the humidity influence the elongation and elasticity of sized yarn?
4. Why the sizing materials and yarn should have the same elongation and elasticity?

Lesson Twenty-Nine

语言微课堂

Weaving

In order to interlace warp and weft threads to produce fabric on any type of weaving machine, three operations are necessary:

(1) Shedding: separating the warp threads into two layers to form a shed.

(2) Picking: passing the weft thread, which traverses across the fabric, through the shed.

(3) Beating-up: pushing the newly inserted weft, known as the pick, into the already woven fabric at a point known as the fell.

These three operations are often called the primary motions of weaving and must occur in a given sequence, but their precise timing in relation to one another is also of extreme importance and will be considered in detail.

Two additional operations are essential if weaving is to be continuous:

(4) Warp control (or let-off): this motion delivers warp to the weaving area at the required rate and at a suitable constant tension by unwinding it from a flanged tube known as the weaver's beam.

(5) Cloth control (or take-up): this motion withdraws fabric from the weaving area at the constant rate that will give the required pick-spacing and then winds it onto a roller.

The yarn from the weaver's beam passes round the back rest and comes forward through the drop wires of the warp stop motion to the heald, which is responsible for separating the warp sheet for the purpose of shed formation. It then passes through the reed, which holds the threads at uniform spacing and is also responsible for beating-up the weft thread that has been left in the triangular warp shed formed by the two warp sheets and the reed. Then, fabric passes over the front rest, round the take-up roller, and onto the cloth roller (Fig.29.1).

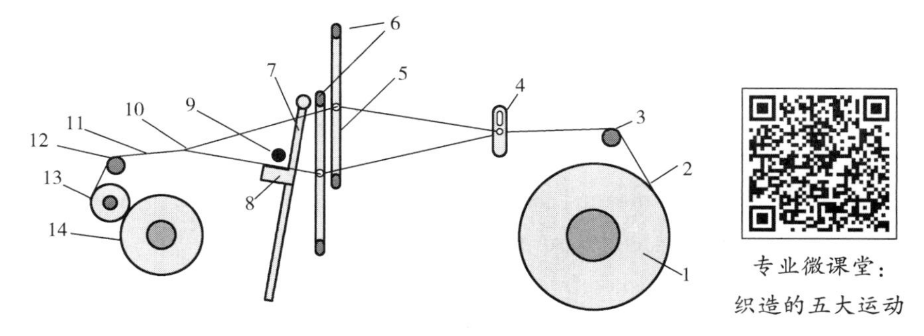

Fig.29.1 Yarn arrangement in loom

1—weaver's beam 2—warps 3—back rest 4—drop wires 5—healds 6—harness frames 7—reed 8—lay 9—weft 10—fell 11—fabric 12—front rest 13—take-up roller 14—cloth roller

Consider the simplest plain shuttle loom on which only two harness frames are used for the shedding operation. The shedding cycle for plain weave (which is the simplest weave) will repeat on two picks or two weaving cycles. In one weaving cycle, the front harness is in the top position and the back harness is in the bottom position. In the next weaving cycle the harnesses change position to produce the pattern, and this sequence is repeated. When the harnesses are in their maximum up and down positions, the shed is said to be fully opened. At the time the two sheets of warp become level, the shed is said to be closed. The moment at which the shed closes is called the shed crossing point. The timing of the crossing within the weaving cycle has a great effect on weaving and the woven fabric(Fig.29.2).

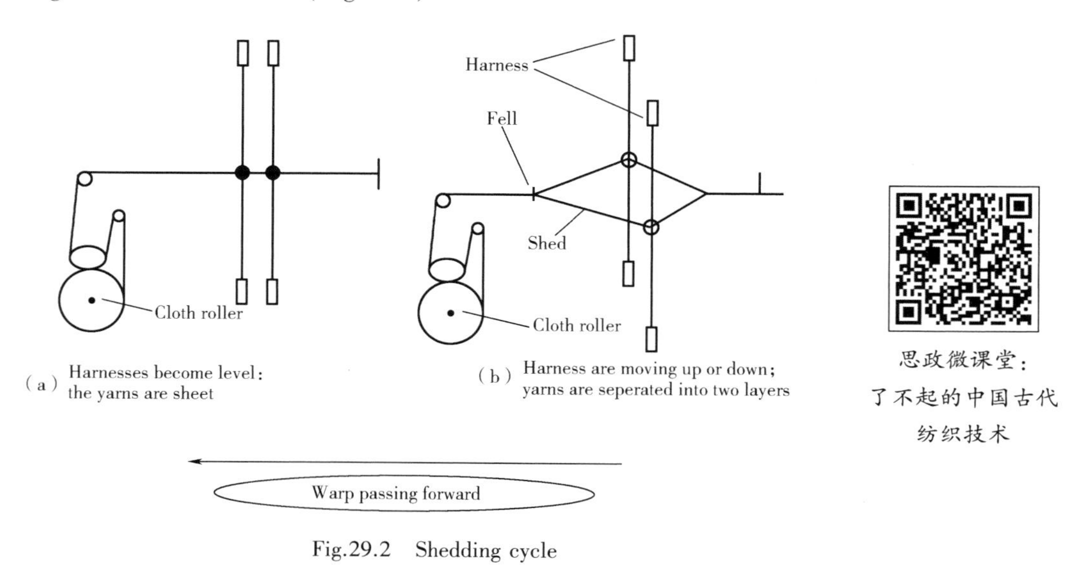

Fig.29.2　Shedding cycle

The shed has to be sufficiently opened and remain open long enough for the shuttle passage. The speed at which the shuttle is projected through the shed is very important in determining the loom speed. It takes a certain period of time to accelerate the shuttle from rest to a speed of some 12m/s in a distance less than 25cm. Also it requires a finite time to stop the shuttle on the other side of the loom. Normally, the shed starts to change long before the shuttle is completely out of the shed. In most cases it is possible to allow the shuttle to enter the shed before the opening of the shed in front of the reed is as large as the shuttle size. Because of this, some friction will exist between the shuttle and the warp yarns near the selvages. Fig.29.3 illustrates the situation when the shuttle enters or leaves the shed.

After the filling is inserted, by whatever means of insertion, it must be incorporated into the body of the fabric. This is the purpose of beat-up.

Beat-up is accomplished by the use of wire grate called a reed. The warp yarns are threaded through the spaces of the reed called dents. At the beginning of the weaving cycle, the reed is moved backward to allow for filling insertion. After the filling is inserted, the reed moves forward and the wires engage the filling yarn, driving it into the fell of the cloth, the position in the warp

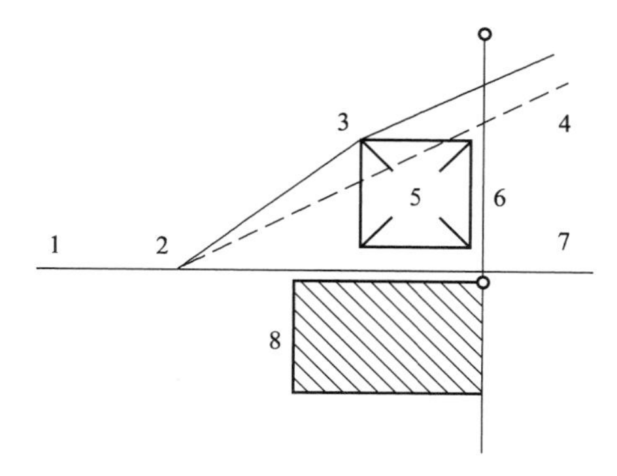

Fig.29.3 Shuttle entering or leaving the shed

1—fabric 2—fell 3—warp deflection 4—upper warp sheet position if there was no shuttle 5—shuttle 6—reed 7—lower warp sheet 8—lay

shed where each pick is beaten. The usual mechanism which gives the reed its motion is known as the lay.

In weaving, the warp and filling tensions are very important factors and have to be controlled. The shedding and beating-up cause cycle tension variations in the warp yarns. On most looms there is a special mechanism to alter the path of yarn to relieve the tension. In its simpler form, the mechanism consists of a cam operated lever to vibrate the back roll, called the whip roll.

To make the process of weaving continuous, the warp yarn must be continuously supplied and the woven fabric must be removed.

The warp let-off motion allows the feeding of warp yarn under uniform tension and at the proper rate. In regular loom, let-off systems fall into three major categories.

The simplest let-off system is known as negative let-off. This system merely consists of a brake continually applied to the warp beam to retard the motion of the beam. The force required to turn the beam is supplied by the tension in the warp yarns.

The second category is known as adjustable let-off. In this system the loom physically drives the warp beam through a gear or servo motor system to feed warp yarns at a predetermined rate. The let-off may be activated intermittently or continuously.

The third system is known as positive let-off. This system is primarily used to produce fabrics in which some of the warp yarns must be fed at different rates, for example, terry cloth.

There are two basic take-up mechanisms, the direct and indirect. In the direct the fabric is taken up and stored by wrapping it around a single beam called the cloth roll. To remove the fabric, i. e. , to doff the cloth roll, the loom must be stopped.

The other method of fabric take-up is to have a set of rolls do the actual fabric take-up and have the cloth roll merely serve to wind the fabric. This system is known as indirect fabric take-up.

It is necessary to hold the fabric at the fell of the cloth to insure that the reed beats the filling across the width of the fabric as nearly parallel as possible. This lateral fabric control is the function of the temples. The temples are simply mechanical “fingers” which maintain the fabric at full width at the fell of the cloth.

New Words and Expressions

interlace　交织
shed　梭口
shedding　开口
picking　投梭，引纬
beating-up　打纬
letting-off　送经
taking-up　卷取
flanged　带凸缘的，用法兰连接的，折边的
pick　纬纱
fell　织口
pick-spacing　纬纱间距
back rest　后梁
warp stop motion　经纱断头停车装置
heald　综框
temple　边撑
front rest　胸梁
take-up roller　刺毛辊
cloth roller　卷布辊
plain shuttle loom　平纹织物织机
harness frame　综框
servo motor　伺服电动机
weaving cycle　织造循环
shedding cycle　开口循环
plain weave　平纹织物
pattern　花型
shed crossing point　综平点
selvage　布边
back roll　后梁
lay　筘座
dent　筘齿、筘隙
negative let-off　消极式送经
semi-positive let-off　半积极式送经
positive let-off　积极式送经
direct take-up　直接卷取
indirect take-up　间接卷取
cloth roll　卷布辊

Notes to the Text

1. pushing the newly inserted weft, known as the pick：在中文中“weft”和“pick”的意思均为纬纱，故“known as the pick”可不必翻译。
2. in relation to one another：指三者之间的相互关系。
3. Temples hold the cloth firm：“firm”意思为“牢固地”，做副词。
4. Take-up roller：指的是刺毛辊，又称卷取辊，较常用的表达方式是“sand roller”。
5. back rest：其另外的表达方式为“whip-roll, back roll”。
6. front rest：其另外的表达方式为“breast roller, breast rest, front roll”。
7. At the time the two sheets of warp become level：“level”的意思是“综框平齐”。
8. shed crossing point：直译是“梭口交错点”，这不是专业术语；它表示的是综框平齐的那一瞬间，即综平点。
9. The speed at which the shuttle is projected through the shed：其中的“the speed”指的是梭子的初始速度。
10. cycle tension variations：指的是一个织造循环内（即主轴转一转时）的张力波动。
11. the path of the yarn：指纱线通过的路径。

12. selvage：布边，是美国拼法，英国拼法为 selvedge。
13. In its simpler form, the mechanism consists of a cam operated lever to vibrate the back roll, called the whip roll：由于"back roll""whip roll"在中文中均指后梁，故"called the whip roll"不必译出。
14. some others repeat their movements every two or more weaving cycles：这里"more weaving cycles"指的是，当织造非平纹织物时，开口凸轮轴的运动状况，例如织造$\frac{2}{1}$斜纹织物时，主轴转三转，即三个织造循环时，开口凸轮轴转过一转。
15. For weaves where it is necessary to use more than two harnesses, an additional shaft is used：讲的是，由于织物组织的需要而增加综框数，并不包括为了降低综框上的纱线密度而使用较多综框的情况。例如织造平纹织物时，有时需要 4 页综或 6 页综等。
16. lay：筘座，另外的拼法有"slay""sley"。
17. The warp yarns are threaded through the spaces of the reed called dents："dent"的含义有两个：筘齿，筘隙。在此应译为"筘隙"。
18. Let-off systems fall into one of three major categories：如果直译为"送经系统是三种方式中的一种"，则显得文理不通。可译为"送经系统有三种"。
19. 在本文中"direct take-up"指织物绕过刺毛辊后，便卷在卷布辊上；"indirect take-up"指织物绕过刺毛辊，经过一对导布辊后，再卷在卷布辊上。

Questions to the Text

1. What are the five main motions in a weaving machine?
2. Prepare a talk on the warp passage through the loom, from the back rest to the cloth roller.
3. How do the shedding, picking and beating-up motions work when producing plain weave?
4. Why there are frictions between the shuttle and the warp sheets?
5. How do the three let-off mechanisms work?
6. What are the differences between the direct and indirect take-up motions?
7. What are the functions of the temple?

Lesson Thirty

Shedding*

语言微课堂

A shuttle loom has two cam operated picking mechanisms, one on either side, each of which operates every other weaving cycle. The lay completes one cycle of its movement in one weaving cycle. Clearly, some of the mechanisms repeat their movements every weaving cycle and some others repeat their movements every two or more weaving cycles. Thus a normal loom has two shafts. The main shaft (crankshaft or top shaft) drives the lay mechanism and is geared to another shaft which operates at half of the speed of the crankshaft. This second shaft is known as the cam shaft, and drives the picking and shedding mechanisms. In the case of a plain loom the shedding cams can be fixed directly on the cam shaft. For weaves where it is necessary to use more than two harnesses, an additional shaft is used, called the auxiliary cam shaft. The speed of the auxiliary cam shaft depends on the number of picks in the repeat of the fabric design. Fig.30.1 gives the arrangement of the shafts in the loom.

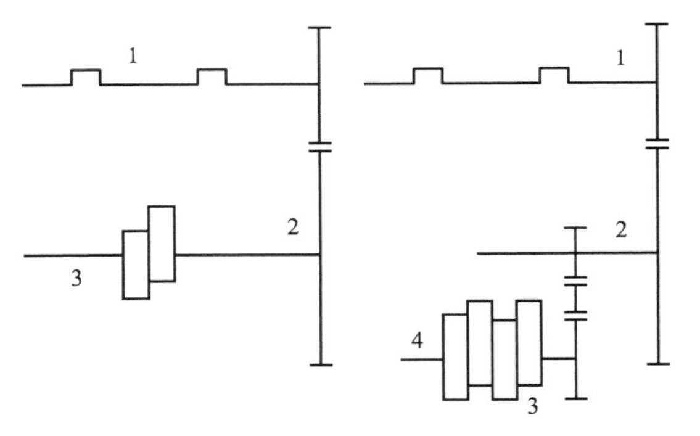

Fig.30.1 The shafts of a loom

1—crank shaft 2—cam shaft 3—cams 4—auxiliary cam shaft

专业微课堂：
开口

If the shed is unclear, it is possible for the shuttle to cause many end breakages and it is usually desirable to keep the front shed angle the same for all ends in order to produce a clear shed. This requires an adjustment to the movements of the various heddles unless they can all exist in a plane at a fixed distance from the fell of the cloth. This is an impossible condition for a normal loom because the harnesses must be at different distances from the fell in order to permit their independent movement. To maintain a clear shed, the front shed angle has to be kept constant and to do this, each harness must be given a slightly different lift from its neighbors; in fact, the lift must be proportionate to the distance of the harness from the fell. Thus the differences depend upon how closely the harnesses can be packed; if there are many harnesses, the differences can be large.

The timing of the opening of the fully open shed is important. Various systems have been designed to compensate for the change in warp length in shedding and so minimize the variations in warp tension. These systems operate on all warp ends, and if some ends are left in position whilst others are moved, the compensation will act on all other yarns, this may cause some ends to go slack and some to be over tensioned. The slackness might enable entanglements to form and the high tensions could lead to end breaks, the variations in tension would also affect the fabric structure.

The shedding mechanisms can fall into three categories, cam, dobby and Jacquard. The lift of the heddles can be achieved by using cam or dobby operated harnesses. The Jacquard system can achieve the lift of the single warp yarn.

In general, the cam system can weave the simple patterns or basic weaves. But most cam systems tend to be inaccessible and it is sometimes inconvenient and time consuming to change the weave, especially with complex fabric designs. With such complex weaves, a change can involve replacement of some or all of the cams, which is a fairly lengthy process.

Dobby designs or patterns are characterized by small figures such as dots, geometric designs, and small floral patterns that have been woven into the fabric. These designs are produced by the combination of two or more basic weaves, and the loom may have up to 32 harnesses to control the shedding operation. The designs are woven by dobby loom. And they are produced by a dobby pattern chain, which may be made of wood with metal pins or perforated metal, paper or plastic rolls. The pattern chain controls the harnesses and determines which are to be raised or lowered in order to produce the desired pattern. Each cross bar or row of holes in the roll controls a row of the pattern and mechanically determines which warp yarns (harnesses) will be raised and which lowered. Examples of fabrics produced by dobby weaving are piqué, waffle cloth and shirting madras. Piqué may utilize heavy yarns, called stuffer yarns, to produce an accentuated pattern.

Fabrics with extremely complicated woven designs are manufactured using Jacquard shedding attachments. The major advantage of the Jacquard attachment is its ability to control each individual warp yarn instead of a series as in regular harness looms. Fabric designs can be changed merely by inserting a series of punched cards into the pattern chain. This separate yarn control provides great freedom for the fabric designer because large, intricate motifs can be transferred to fabric. However, the Jacquard heads are expensive and the production rate is very low and consequently these looms are used only for complex weaves. Jacquard fabrics are used for home furnishings, apparel, domestics such as table coverings and napkins, and other elaborate and decorative fabrics.

Jacquard system is among the older types of shedding mechanisms. Until recently all Jacquard designs were created using a set of pattern cards. The pattern desired is transferred to a series of perforated cards or a roll; one card is required for each shedding and placing of a filling pick. The holes in the card permit the lifting of hooks on the machine that determine which warp yarns are to be raised in forming the shed. Each card stops on the cylinder for its particular pick, the pick is placed, the card moves on, and a new card takes its place. This continues until all cards are used and one repeat of the pattern is formed; the process is then repeated for as long as needed.

Many Jacquard attachments in use today are controlled by computer rather than cards. The process is basically the same, but the computer identifies which yarns are to be raised to form the desired shed. Complicated patterns can be developed using various kinds of software that permit the drawing of designs directly onto monitors that are then converted into the pattern tape for controlling the shedding of the warp yarns. Looms using computer controls appear very similar except that the cards are missing and the controls are housed in the upper-level control box.

New Words and Expressions

crankshaft　曲拐轴
cam shaft　织机中轴，织机踏盘轴，织机凸轮轴
shedding cam　开口踏盘，开口凸轮
auxiliary cam shaft　辅助踏盘轴，辅助凸轮轴
heddle　综丝
front shed angle　前部梭口角
clear shed　清晰梭口
punched cards　纹板
pattern chain　纹链
complex weave　复杂组织
Jacquard head　提花龙头
regular figure　规则花纹
patterned fabric　花纹织物
floral pattern　花卉图案
basic weave　基础组织
dobby weaves　多臂组织(指多臂织机可以织造的织物组织，小花型组织)
piqué　凹凸组织
waffle cloth　蜂窝纹布，方格纹布
shirting madras　衬衫布
stuffer yarn　衬垫纱线，填充纱线
hook　提花机竖钩
motif　花纹图案
domestics　家用织物
napkin　餐巾

Notes to the Text

1. number of picks in the repeat of the fabric design：指完全织物组织的循环纬纱数。
2. if some ends are left in position whilst others are moved, the compensation will act on all other yarns：句中的“left in position”指经纱处于原来的位置，没有运动。
3. Thus the differences depend upon how closely the harnesses can be packed：“the differences”指“综框动程的差异”；“packed”可译为“排列、安排”。
4. Most cam systems tend to be inaccessible：多数凸轮装置不便于修理或更换(“inaccessible”意为“不易接近的”)。
5. This separate yarn control provides：句子中的“controls”是名词，意思为控制机构。
6. instead of a series as in regular harness looms：“regular harness looms”指的是使用综框的常规织机。
7. one card is required for each shedding and placing of a filling pick：“placing of a filling pick”意思是“纬纱的引入”。

8. that permit the drawing of designs directly onto monitors: "drawing of designs" 意思是"绘制的图案"。

Questions to the Text

1. How to fix the shedding cams for various weaves?
2. How to get a clear shed?
3. How to control the warp tension in a loom?
4. Compare the three main shedding mechanisms, the cam, the dobby and the Jacquard.
5. How does the dobby control the harnesses or warps?
6. What's the major advantage of the Jacquard looms?
7. How to control the individual warp in the Jacquard loom?
8. How do you know about the computer-controlled Jacquard mechanisms?

Lesson Thirty-One

Shuttleless Looms

语言微课堂

According to the means of weft insertion, the shuttleless looms include:

(1) Projectile loom.

(2) Rapier loom.

(3) Air jet loom.

(4) Water jet loom.

思政微课堂：
延续千年的
中国织造技术

Projectile loom.

The projectile loom uses a small projectile to carry the filling yarn through the shed. In reality, the shuttle is actually a projectile but the distinctive feature of shuttleless looms that the filling package be exterior to the filling carrier. The projectile loom uses a projectile with grippers to carry the yarn through the shed. This projectile may be picked from both ends or from one end. If picking is from one end, there must be a method to carry the projectile back to the other side of the loom, and a chain return mechanism is the best choice.

Because the mass of the projectile is much less than that of the conventional shuttle, the forces needed to accelerate it are less and the picking mechanism can be lighter; this in turn reduces the total mass to be accelerated and makes it possible to use new system. Also, because the mass is low, the speed can be increased. Thus these looms can be run faster than conventional shuttle looms. Also the acceleration of the projectile can exceed that of a shuttle by a factor of about 7.

The normal projectile is rather short and if it were to meet a substantial obstruction in passing across the loom, it could quite easily be deflected; at worst it could fly out of the loom, which could be very dangerous. Also the collection of the projectile after it has completed its task of carrying the filling is made more difficult if it does not follow an accurate path. For these reasons, a series of guides are used to constrain the projectile as it passes across the loom, see Fig.31.1. To make this possible, the projectile guides or gripper guides must protrude through the warp sheets.

Rapier loom

Rapier looms are made in a variety of types. The rapiers may be required to extend across the full width of the warp, in which case they will be of a rigid construction, or alternatively two rapiers may enter the shed from opposite sides of the loom and transfer the weft from one rapier head to the other at a point near the center of the loom. In the later case, the rapiers may be either

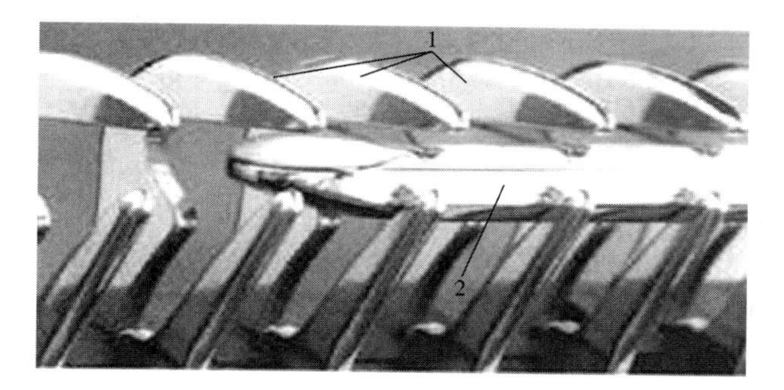

专业微课堂：
织机

Fig.31.1 Projectile guides and projectile
1—projectile guides 2—projectile

rigid or flexible. The future of this method of weft insertion would appear to lie mainly in the field of multi-color work, since their rates of weft insertion are generally only comparable with those of conventional automatic looms.

A rapier is merely a device made of metal or a composite material with an attachment on the end to carry the filling yarn through the shed. The simplest rapier loom is known as the single rigid rapier. This type of loom consists of one rapier which carries the yarn through the shed. In most instances this loom is not as fast as a shuttle loom because the rapier must be withdrawn from the shed prior to beat-up and, during this withdrawal, no filling is carried. There are also space limitations, since the single rapier must be at least as long as the loom width.

To overcome the potential space problem, the double flexible rapier was developed. This loom consists of two rapiers, one called a "giver" and the other a "taker". The giver picks up the filling yarn and carries it to the center of the shed. Meanwhile, the taker moves to the center of the shed, picks up the yarn from the giver, and carries it the remaining distance. Now each rapier covers only half the distance and therefore almost twice number of picks per unit time can be inserted as compared to the single rapier. The rapiers in this case are no longer rigid bars but flexible tapes. Since the tapes are flexible, they can be bent so as to fit under the loom, thus conserving space.

Air jet loom

In an air jet loom, high pressure air carries the filling yarn through the shed. The major drawback to this type of loom is that the air diffuses quite rapidly so the width of the fabric is limited. In practice, it is not possible to project a filling more than about 1.5m with a simple air jet of reasonable size and power consumption. Therefore, the width of the loom is limited too. Modern air jet looms use air guides or air guides with auxiliary jets to extend the distance that the yarn can be carried.

A blast of air would seem to be an effective way of inserting the filling, but to get enough traction on the filling yarn it is necessary to use very high air velocities. The air just emerging from the nozzle is highly energetic and, as it moves away from the nozzle, it entrains some of the surrounding air which tends to slow the mass down. Thus as the moving mass of air moves away, it grows larger and become slower. The consequence is that the loom is noisy and consume considerable amount of energy. This increases the cost of weaving and tends to make air jet looms less attractive.

One way of improving the performance of an air jet loom is to use a device to prevent the air jet from breaking up so quickly. A series of orifices, with slots to permit the removal of the filling at beat-up, can be placed along the filling axis. These act as a sort of porous tube and tend to improve the axial air velocity. They also reduce the turbulence so that there is less disturbance to the filling. Such a system permits an increased productivity by allowing a higher loom speed or wider loom or both to be used. The orifices are sometimes called confusors, more recently most of them have been replaced by tunnel reeds.

Water jet loom

The technique of inserting weft in a fluid jet achieves weft-insertion rates comparable with those achieved by the Sulzer weaving machine, but there are width limitations. Water jet looms have been shown to be most economical for producing certain types of plain continuous filament fabric, especially when they are produced from hydrophobic synthetic fiber fabrics.

In a water jet loom, the filling yarn is carried by a high pressure jet of water. Since water diffuses much more slowly than air, wider fabrics can be made with water jet looms without using guides. However, because water is the carrier, there are restrictions as to yarn types and pre-weaving processing, such as slashing. The droplets spread in such a way as to wet much of the warp, thus a sized warp containing a water soluble adhesive can be adversely affected. Because of this, water jet weaving is usually restricted to hydrophobic filament yarn, but there is some hope that it might become economically feasible to weave staple yarns on these looms.

A water jet is more coherent than an air jet. It does not break up so easily, and the propulsive zone is elongated, making it much more effective. It is effective in terms of energy requirements, it is quiet and, when the jet does break up, it goes into droplets which create very little turbulence to disturb the filling.

Two main reasons for the efficiency of the water jet loom are that there are no varying lateral forces to cause the filling to contort, and the moving element is more massive because it is wet. Thus there is less chance of fault due to contact with the warp. So this loom have very high rate of weft insertion.

New Words and Expressions

projectile loom　片梭织机
gripper loom　剑杆织机，片梭织机
rapier loom　剑杆织机
water jet loom　喷水织机
air jet loom　喷气织机
filling carrier　载纬器
grippers　夹子，梭夹
chain return mechanism　链条式片梭返回装置
collection　接收
projectile guides　导梭片
gripper guide　导梭片

conventional loom 常规织机，有梭织机
collection 接收
composite material 复合材料
rapier head 剑头
flexible rapier 挠性剑杆
giver 递纬剑，送纬剑
taker 接纬剑
nozzle 喷嘴
diffuse 扩散，使分散
auxiliary jet 辅助喷嘴
turbulence 湍流
confusor 管道片(喷气织机防气流扩散的构件)
tunnel reed 异型筘(喷气织机防气流扩散的构件)
Sulzer 苏尔寿(瑞士公司)
propulsive zone 推进区，牵引区
lateral force 侧向力
contort 歪斜
rate of weft insertion 入纬率(表征织造产量的指标)

Notes to the Text

1. means：方法，手段。此词单复数同形。例如：a means to an end：达到目的的方法；by all means：尽一切办法。
2. gripper shuttle，gripper，projectile：都可表示片梭，但常用的是“projectile”。
3. The normal projectile is rather short and if it were to meet a substantial obstruction：“were to meet”使用的是虚拟语气，表示实际上不可能发生。
4. To make this possible，the projectile guides or gripper guides must protrude through the warp sheets：“warp sheets”指“下层经纱”。
5. The rapiers may be required to extend across the full width of the warp，in which case they will be of a rigid construction：句子讲的是单剑杆织机。“rigid construction”是指“刚性结构”。
6. The future of this method of weft insertion would appear to lie mainly in the field of multi-color work：“to lie mainly”意思是“主要在于”；“multi-color work”指的是“多色织物”。
7. are generally only comparable：一般只能与……相比，一般和……差不多。
8. plain continuous filament fabric：简单的长丝织物。
9. taker：接纬剑，另外一种表示方法为“receiver”。
10. carries it the remaining distance：句子中“the remaining distance”指送纬剑没有走完的梭口中的一段距离。
11. auxiliary jet：辅助喷嘴，也可称为“auxiliary nozzle，sub nozzle”。
12. but to get enough traction on the filling yarn：为了给纬纱足够的牵引力。
13. it entrains some of the surrounding air which tends to slow the mass down：句中的“to slow the mass down”指使运动着的空气降低速度。
14. tunnel reed：异型筘，另一种表达方式是“profiled reed”。
15. A water jet is more coherent than an air jet：句中的“coherent”意思是“黏附的、不易扩散的”。
16. Thus there is less chance of fault due to contact with the warp：句子中“contact with the

warp"指飞行的纬纱与经纱的接触。

Questions to the Text

1. How many types the shuttleless looms include?
2. What is the projectile loom?
3. How to keep the flying projectile in an accurate path?
4. What are the disadvantages of the single rapier loom?
5. Describe the types of rapier looms you know.
6. How to improve the speed and width of air jet looms?
7. What would happen to the air jet after it is out of the nozzle?
8. Describe the reason that the water jet loom must use the hydrophobic materials.

Lesson Thirty-Two

Basic Weaves

语言微课堂

Plain weaves

The plain weave is the simplest form of yarn interlacing and consists of the alternate shedding of warp yarns to provide a fabric in which each filling yarn passes one warp and under the adjacent warp. An adjacent filling yarn reverses the interlacing so that it passes over the warp that on the top of the preceding filling yarn and under the warp that lay under the preceding filling yarn.

Many woven fabrics, say 70%, are constructed using the plain weave. Unless colors have been printed onto the surface, plain weave fabrics are generally considered to be reversible. In making plain weave fabrics, yarns can be packed loosely or tightly together, and the number of warp yarns may equal to the number of filling yarns in each 10cm or may vary considerably. Yarns can vary in diameter and type to produce interesting fabrics. When the number of warp yarns per 10cm is approximately the same as the number of filling yarns per 10cm, the thread count, sometimes called fabric count, is considered balanced; when the number of yarns per 10cm differs considerably between warp and filling, the fabric count is unbalanced.

The plain weave is comparatively inexpensive to produce. A variety of designs can be made using various coloring and finishing processes on the finished fabric. In addition, some of the most durable fabrics are manufactured by the plain weave technique. Examples of plain weave fabric include muslin, percale, cheesecloth, chambray, gingham, batiste, organdy, chiffon, shantung and homespun.

Interesting and attractive fabrics can be obtained by utilizing the rib variation of the plain weave. The rib effect is produced by using heavy yarns in the filling or warp direction (usually the filling direction), by grouping yarns in specific areas of the warp or filling, or by having more warp than filling yarns per 10cm.

Many rib fabrics have heavy yarns inserted as picks. Examples of this construction include poplin, faille, bengaline and ottoman.

Twill weaves

The twill technique is characterized by a diagonal line on the face, and often on the back, of the fabric. The face diagonal can vary from a low 14° angle (reclining/flat twill) to a 75° angle (steep twill). An angle of 45° is considered to be a medium diagonal or a regular twill. The regular twill is the most common. The angle of the diagonal is determined by the closeness of the warp ends, the number of yarns used per 10cm, the diameter of the yarns used, and the actual progression forming the repeat.

Twill weave fabrics have a distinctive and attractive appearance. In general, fabrics made by the twill weave interlacing are strong and durable. Twills differ from plain weave in the number of filling picks and warp ends needed to complete a repeat pattern. The simplest twill uses three picks and three warp ends to form the repeat. At least three harnesses are required on looms if a twill weave is to be made.

The warp yarn goes over, or floats over, two filling yarns and under one in the $\frac{2}{1}$ twill. In a regular twill each succeeding float begins one pick higher or lower than the adjacent float. In more complicated twills the progression may vary, but the diagonal effect will remain visible. The number of pick yarns required to complete the twill pattern determines the number of harnesses required for weaving. Some twill patterns may require as many as fifteen harnesses.

Twill fabrics have either a right-hand or left-hand diagonal. If the diagonal moves from the lower left to the upper right of the fabric, it is referred to as a right-hand twill; if it moves from upper left to lower right, it is a left-hand twill. In even twill fabrics, the filling yarns pass over and under the same number of warp yarns, whereas in uneven twills the pick goes over either more or fewer warps than it goes under. Uneven twill fabrics have a right and a wrong side and therefore are not considered reversible. Filling-faced twills are those in which the picks predominate on the face of the fabric; warp-faced twills have a more evident warp on the surface. Even twills have the same number of warp and filling yarns on both the face and back and can be considered reversible. Of course, finishes and coloring methods may make one side the face and prevent reversing the fabric. Fig.32.1 shows various twill weaves.

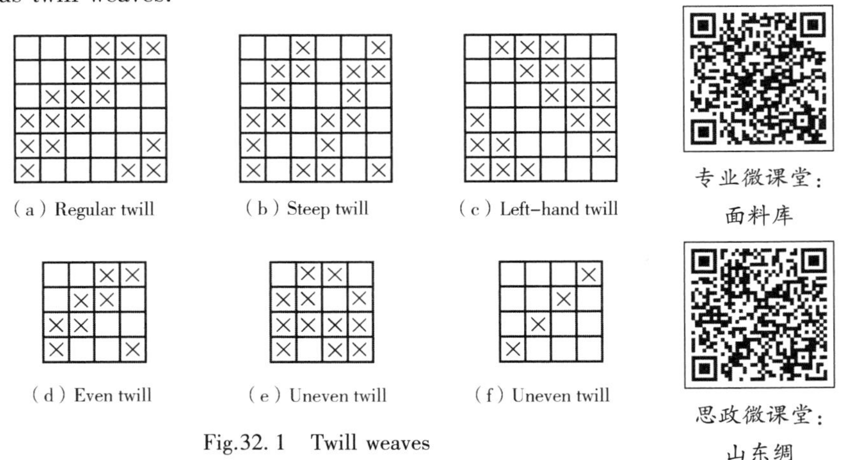

(a) Regular twill (b) Steep twill (c) Left-hand twill

(d) Even twill (e) Uneven twill (f) Uneven twill

Fig.32.1 Twill weaves

Yarns used in making twill weave fabrics frequently are tightly twisted and exhibit good strength. The twill weave permits the packing of yarns closer together because of fewer interlacings. This close packing combined with strong yarns can produce a strong, durable fabric. However, if the yarns are packed too tightly, the fabric will have reduced tearing strength, abrasion resistance and wrinkle recovery.

In addition to appearance and good strength, twills tend to show soil less quickly than plain weave fabrics. However, twills are more expensive to produce than plain weave fabrics because of the more complicated weaving techniques and looms are required.

A frequent variation of the twill weave is the herringbone. In this design, the twill direction is reversed at predetermined intervals to form a series of inverted "V"s.

Examples of fabrics made by the twill weave include denim, jean, gabardine, surah, sharkskin, some flannel fabrics and some tweeds.

Satin and sateen weaves

Satin and sateen weaves are characterized by long floats in the warp-and-weft directions, respectively. These floats are caught under cross yarns at intersections as far apart as possible for the particular construction. At no time do adjacent parallel yarns come in contact with one another at a point of interlacing. This reduces the possibility of a diagonal effect occurring on the face of the fabric. However, some satins do have a faint diagonal effect owing to the low number of yarns used to form the repeat.

In a satin fabric it is the warp ends that float on the surface. A variation of the satin weave in which the filling picks float on the surface is referred to as sateen. Filament fiber yarns are generally used in making satin fabrics, while staple-fiber yarns, often of cotton, are more common in sateen fabrics. There are, however, exceptions; cotton satins as well as filament yarn sateens have been made. Fig.32.2 gives the structures of the sateen and satin.

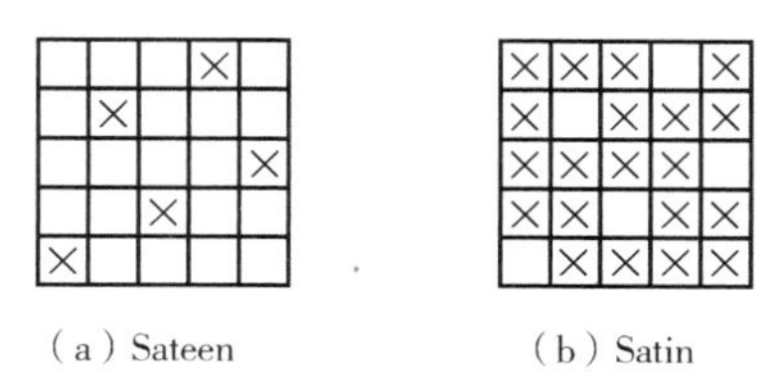

Fig.32.2 Sateen and satin weaves

The long floats of the satin weave create a shiny surface and tend to reflect light easily. This is accentuated if bright filament fiber yarns with a low twist (about 20–40 turns per meter) are used for the floating yarns. Floats in satin and sateen fabrics tend to snag and abrade easily, thus, such fabrics are not as durable as plain or twill weave fabrics.

The length of the floats in satin structures is governed by three factors: the number of harnesses, which is determined by the number of filling yarns involved in forming the repeat; the number of yarns per 10cm, or fabric count; and the yarn size. Of these, the most important are fabric count and yarn size. A satin fabric may have as few as five harnesses to form the repeat (in which case it is called a 5-shaft satin), or it may have many more. The number of yarns per 10cm influences the length of the float along with the number of harnesses. Among fabrics with the same number of yarns required for the repeat, a fabric with a high fabric count will have shorter floats than one with a low fabric count. When fabric count and yarn size are held constant, the length of a float over four yarns will obviously be shorter than a float over seven yarns.

Satin weave fabrics are lustrous and are selected primarily for appearance and smoothness. Satins are frequently used as lining fabrics in coats, suits and jackets, as they make it easy to slip the apparel item on and off over other materials. Satins and sateens have a definite face and back. Variations in satin fabrics can be produced by using highly twisted yarns in the filling to create a crepe-back satin; when the crepe side is intended to be used as the right or face side of the fabric, it is called a satin-back crepe.

Examples of satin weave fabrics include antique satin, slipper satin, crepe-back satin, bridal satin and moleskin.

Words and expressions

thread count 织物经纬密度
fabric count 织物经纬密度
muslin 平纹细布，薄纱织物
percale 高级密织薄纱
cheesecloth 干酪包布
chambray 钱布雷平布，色经白纬平布
gingham 方格色织布
batiste 细薄织物，法国上等细亚麻织物
organdy 蝉翼纱
chiffon 雪纺绸，薄纱
shantung 山东绸
rib fabric 凸条纹织物
homespun 钢花呢，手工纺织呢
poplin 府绸
faille 罗缎，菲尔绸
bengaline 罗缎，孟加拉织品
ottoman 粗横棱纹织物 ottoman［简明英汉词典］（=Turkish）土耳其帝国的，土耳其人的
twill weave 斜纹
reclining twill 缓斜纹
steep twill 急斜纹
regular twill 正则斜纹
progression 飞数
float 浮长线
right-hand twill 右斜纹
left-hand twill 左斜纹
even twill 双面斜纹
uneven twill 单面斜纹
right side 织物正面
wrong side 织物反面
filling-faced twill 纬面斜纹
warp-faced twill 经面斜纹
tearing strength 抗撕裂强度
abrasion resistance 耐磨强度
wrinkle recovery 折皱回复度
herringbone 人字斜纹
jean 粗斜纹棉布，三页细斜纹布
surah 斜纹软绸
sharkskin 鲨皮布，鲨鱼皮革
flannel 法兰绒
sateen 纬面缎纹
5-shaft satin 五枚缎纹
lining fabric 衬里织物
suit 成套衣服
crepe-back satin 绉缎
satin-back crepe 缎背绉
antique satin 仿古缎
slipper satin 鞋面花缎
bridal satin 婚服缎
moleskin 仿鼹鼠皮，鼹鼠皮

Notes to the Text

1. alternate shedding of warp yarn：经纱交替开口。
2. Unless colors have been printed onto the surface, plain weave fabrics are generally considered to be reversible："reversible"指织物正反面结构相同。
3. the thread count, sometimes called fabric count："thread count"和"fabric count"指织物中的纱线密度，也可用"fabric set"表示。
4. by utilizing the rib variation of the plain weave："rib variation"指的是凸条的变化。
5. balanced (fabric)：一般指经纬纱线密度、经纬密度均相同的织物。
6. The face diagonal can vary from a low 14° angle (reclining twill) to a 75° angle (steep twill)："low 14° angle"意思是"较小的14°角"。
7. the actual progression forming the repeat：构成完全组织的实际飞数。
8. whereas in uneven twills the pick goes over either more or fewer warps than it goes under：此句若译为"对于单面斜纹，处于纬纱下面的经纱根数大于处于纬纱上面的经纱根数，或处于纬纱上面的经纱根数大于处于纬纱下面的经纱根数"则显得比较繁琐；可以译为"对于单面斜纹，处于纬纱上面和纬纱下面的经纱根数是不同的"。
9. In this design, the twill direction is reversed at predetermined intervals to form a series of inverted "V"s："the twill direction is reversed"意思是"斜纹的方向被反向布置"。
10. jean：劳动布，牛仔布；"jeans"为牛仔裤。
11. gabardine：华达呢，该词另外的拼法为："gabardeen""gaberdine"。
12. These floats are caught under cross yarns at intersections as far apart as possible for the particular construction：句中的"are caught under cross yarns at intersections"的意思是"在交织点处被另一组纱线压在下面（即被终止）"。
13. Floats in satin and sateen fabrics tend to snag and abrade easily：句中的"snag"指由于摩擦，纤维被勾出织物表面或起毛。
14. 在一般情况下，"satin"代表所有的缎纹组织；在需要区分经面缎纹和纬面缎纹时，它仅代表经面缎纹。

Questions to the Text

1. How is the plain weave produced in a loom?
2. Give the names and simple explanations of the plain weaves you know.
3. How to construct the rib fabrics?
4. How many twill weaves do you know according to various constructions? How are they constructed?
5. Why the yarns in the twill weave can be packed closer than that in the plain weaves?
6. How is the herringbone fabric produced?
7. What are the differences between the twill and plain weaves?
8. What are the special characters of the satin weaves?

9. What are the differences between the satin and sateen weaves?
10. What are the main usages of the satin fabrics?
11. What factors influence the float length of the satin weaves?

Lesson Thirty-Three

Several Typical Weaves*

语言微课堂

Matt/hopsack weaves—plain weave derivatives

Matt weaves are (see Fig.33.1) produced by extending the plain weave structure both warp and weft ways, so that in both directions there are two or more threads working together in the same order. They have 2, 3 or 4 adjacent ends working as one single end, and 2, 3 or 4 picks in a shed. Matt weaves use either square or nearly square constructions.

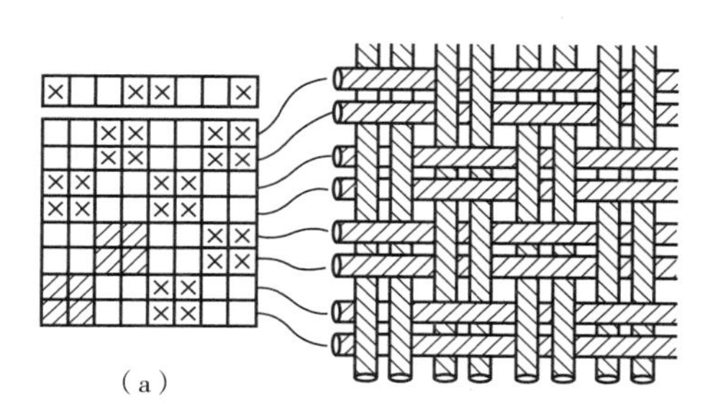

(a)

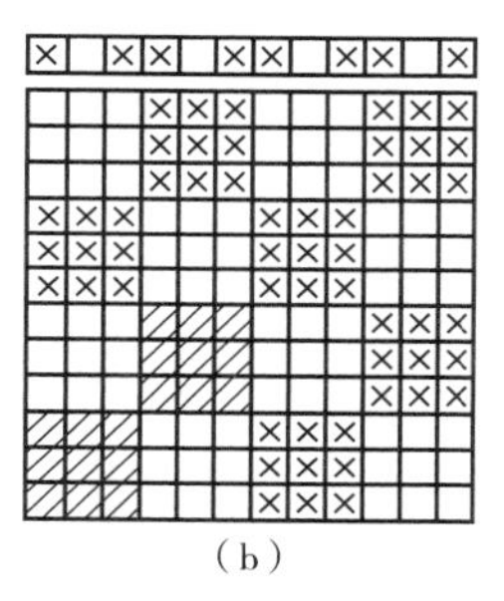

(b)

Fig.33.1 Matt weaves

A simple regular $\frac{2}{2}$ matt will be 25% heavier than a square plain cloth made from same yarns. Due to less number of intersections in the cloth, fabrics are porous or open. The weaves are principally used when we desire to make a fabric that will be very strong, thin and also very dense in the setting of ends and picks per 2.54 centimeter without being too hard and stiff.

The matt weave cloth will also have a much greater resistance to tearing, partly because it has more threads per 2.54 centimeter, but chiefly because the threads are in pairs. When a plain weave cloth tears, the yarns at right angles to the direction of the tear usually break one by one in rapid succession. With $\frac{2}{2}$ matt, the tearing force tends to be borne fairly equally by two threads—the pair which will break next; hence the matt weave cloth has a greater resistance to tearing.

Sailcloth is often made in $\frac{2}{2}$ matt weave because of its high resistance to tearing in both directions. Matt weaves tend to give smooth-surfaced fabrics, but with an appearance rather like that of a plain weave fabric made from much thicker yarns. These characteristics are sometimes useful in dress fabrics, particularly in worsted suits and costume cloths; one can obtain an appearance of coarse texture without the disadvantages of excessive thickness and stiffness. Larger matt weaves such as $\frac{3}{3}$ or $\frac{4}{4}$ give a small, neat, chess-board appearance in the fabric. They are seldom used in their simple form because they require very high cover factors to make a firm cloth. Matt weaves are used for shirtings, canvas, sailcloth filter cloths and also for heavy "ducks" such as conveyor belts etc.

Pointed and diamonds twills

These are characterized by the sharp termination of Z twill line and S twill line at reversal point and are mainly employed in furnishing. Normally the weave repeats on double the number of basic weave. Point end is obtained by arranging the $N+1$th thread same as $N-1$th thread, as shown in Fig.33.2.

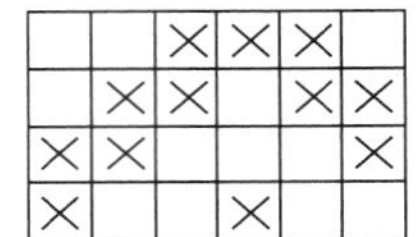

Fig.33.2 An example of point twill

专业微课堂：面料库

Following is the procedure for constructing point twills. Pointed twills may be developed using either warp-faced or weft-faced or equi-faced basic twills.

Diamonds can be developed by employing a vertical waved twill or zig-zag as the lifting plan in conjunction with the point draft, e. g. , two pointed twills are combined such that one is below the other which is tilted as shown in Fig. 33. 3. If a point of waved draft looming is used in conjunction with a waved twill for a peg plan, we produce in the weave plan a series of diagonal lines crossing one another and forming diamond shapes.

The method is illustrated at (a)(b)(c), and (d) in Fig. 33. 3 in which (a) shows a $\frac{1}{3}$ twill that is arranged at(a) as a horizontal waved twill in the order of 1, 2, 3, 4, 3, 2, while(c) represents the same twill arranged to zig-zag vertically in the order of 1, 2, 3, 4, 3, 2 (two repeats are given in each direction). If(b) be taken as(a) draft with(c) as the lifting plan, the small diamond design given at(d) will result.

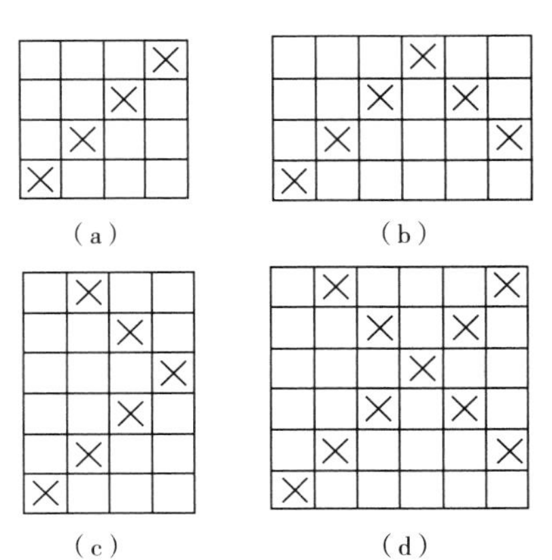

Fig.33.3 Diamond twill

Honeycomb weaves

The characteristic feature of honeycomb (HC)

weave is the presence of alternate raised and sunken diamond shaped areas giving the effect of honeycomb. Honeycombs are characterized by point or V-draft as compared to special ordinary draft of huck-a-backs. In honeycomb weaves, repeat size and the longest float (always odd number) of the design and the number of heald shafts are related. The arrangement is such that the structure results in characteristic point draft. Honeycomb towels are the second class of toweling fabrics used to a greater extent and are made from coarser count like huck-back fabrics, honeycombs are produced by employing point draft unlike special draft as is the case of huck-a-back.

Generally, all HCs are constructed by marking one left and right twill lines and in some cases from left to right only one twill line is marked and from left to right two rows of twill line are marked. Whatever it may be, HCs are observed to be less firm except at the center where plain weave is formed. Therefore, sometimes to overcome this problem, it is customary to add double row of twill lines on both left and right hand sides. These weaves are a distinct class of diamond structures in which the ends and picks are made to lie in different horizontal planes in the fabric, and thus produce ridges and hollows in regular order, giving to the cloth a cell-like appearance. Both warp and weft float freely on face and back side of fabrics. The fabrics are generally woven with fairly coarse yarns and appear thick and spongy, and since they possess long floats of warp and weft, they readily absorb moisture and are suitable for use as towels. When fabrics are made with long floats of warp and weft in close proximity to one another they tend to become thick and have rough surfaces, and this is notably the case if some tight weaving portions are introduced near to the longer floats of warp and weft. In present day these weaves are used for ladies' handkerchiefs when produced in finer counts.

Honeycomb weaves are also used for making bed-spreads and quiltings, and to a more limited extent for dress and coating materials because the weave effect is of a decorative nature.

In most cases these can be woven in point drafts, and a method of constructing the designs on this principle is illustrated at (a) (b), and (c) in Fig. 33.4. A point draft is indicated on the required number of healds—in this case, five, as shown at(a); then the marks are reversed, as indicated at(b). Afterwards, one of the diamond spaces is filled in while the other is left blank, as represented at(c).

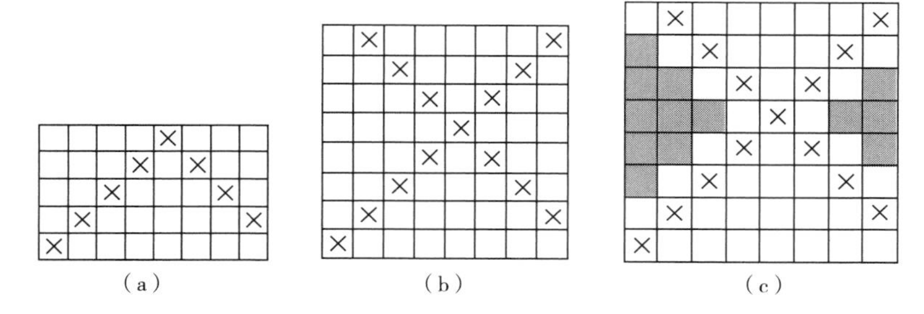

Fig.33.4 The construction of a honeycomb weave

Words and expressions

matt　方平组织
hopsack　方平组织
huck　浮松
proximity　接近
zig-zag　Z 字形
lifting plan　提综图，纹板图，也叫 peg plan
draft plan　穿综图，也叫 drawing plan
heavy “ducks”　重型机械
handkerchiefs　手帕
quilting　被子

Notes to the Text

1. pointed and diamond twill：山形斜纹与菱形斜纹，“pointed”指斜纹线转变的那一点，转变点，故可译为山形斜纹或人字形斜纹，“diamond”，本意指钻石，西方国家有的窗户为钻石形，即菱形。
2. Honeycombs are characterized by point or V-draft as compared to special ordinary draft of huck-a-backs：与普通的浮松织纹图案比，蜂巢组织有人字型或 V 型的花纹特点。“huck-a-back” 源自“hucksters”，小商贩，指具有 13 世纪小商贩售卖浮松麻类织物特征的织物，常随身背在背上卖。“Huck”指一种表面粗糙疏松的棉或亚麻布。

Questions to the Text

1. Why matt weave cloth are better to resistance to tear?
2. How many the methods of developing pointed twills?
3. How to form a honeycomb-like fabric?

Lesson Thirty-Four

Knitting

语言微课堂

Introduction

Knitting is one of the important techniques to form textile structures with intermeshed loops made of curved yarns by the needles. It evolved from hand knitting using simple pins. The Englishman William Lee invented the first stocking hand frame in 1589. Though Lee' s original frame was undoubtedly crude, and knitted poor quality woolen stockings with a gauge of only 8 needles per inch, the principles of it continue to be employed in modern knitting machines.

Unlike weaving, which requires two yarn sets, knitting is possible using only a single set of yarns. The set may consist of a single yarn (weft knitting) or a single group of yarns (warp knitting).

In weft knitting, the loops of yarn could be formed by a single weft yarn. The loops are formed, more or less, across the width of the fabric usually with horizontal rows of loops, or courses.

In warp knitting, all of the loops making up a single course are formed simultaneously. Thus the lengths of each vertical column of loops, the wales, increase at the same time.

The knitted loop structure

Knitted structures are progressively built-up from row after row of intermeshed loops. A stitch is the basic structural unit in both weft and warp knitted fabrics. The stitch in weft knitted structure consists of a head (H), two legs or side limbs (L) and two sinker loops (S), as shown in Fig.34.1.

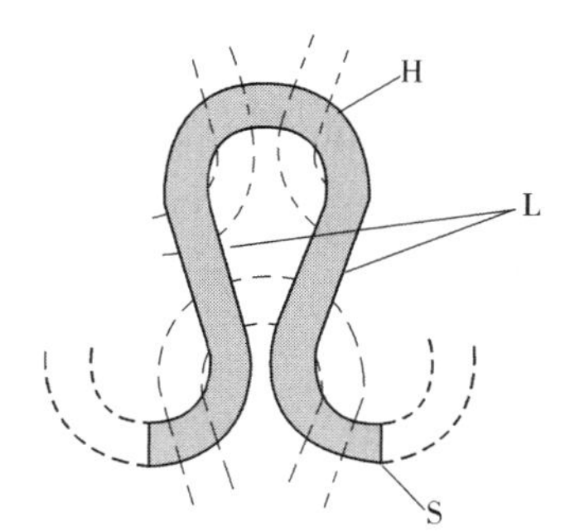

Fig.34.1 Intermeshing points of a weft knitted loop

专业微课堂：针织过程

思政微课堂：中国针织工业

The head and two legs are called the needle loop (H+L in Fig.34.1), which is the common part in both weft and warp knitted loop structure. Along the base of each leg it is the sinker loop. The sinker loop (S in Fig.34.1) is the piece of yarn that joins one weft knitted needle loop to the next. The terms "sinker loop" and "needle loop" are convenient descriptive terms but their precise limits within the same loop length are impossible to exactly define.

Loops are termed "laps" in warp knitting because the warp guides lap their yarn around the needles in order to form the loop structure. A warp knitted lap consists of the overlap (O) and underlap (U), and the former is similar as that in weft knitted loop, which also could be formed by a head (H) and two legs or side limbs (L). The loops (overlaps) may be open (Fig.34.2) or closed (Fig.34.3).

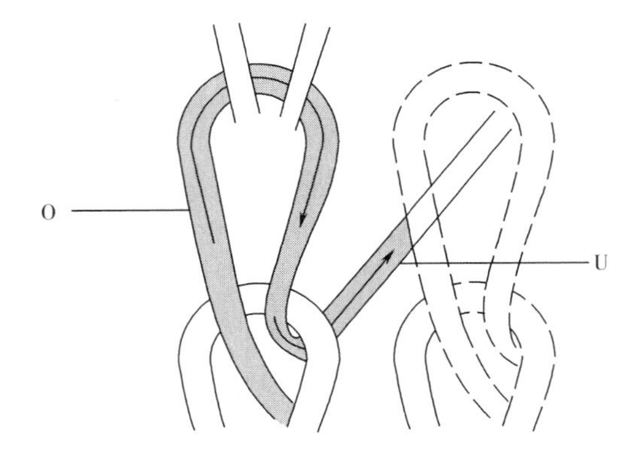

Fig.34.2 Intermeshed open laps of a warp knitted structure

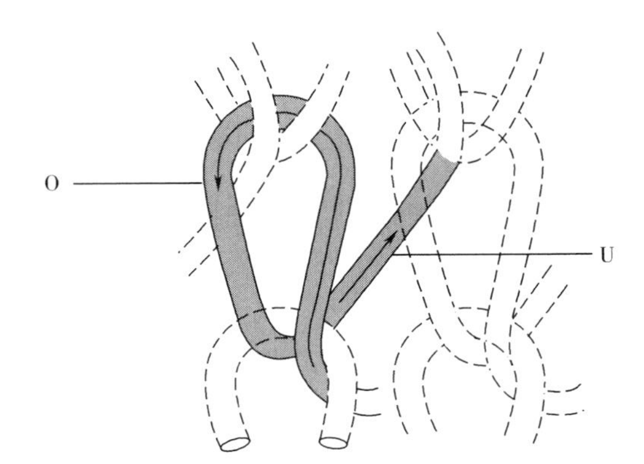

Fig.34.3 Intermeshed closed laps of a warp knitted structure

The overlap (O in Fig.34.2 and Fig.34.3) is a shog, usually across one needle hook, by a warp guide (at the back of a single needle bar machine) which forms the warp yarn into the head of a needle loop. Every needle on a conventional warp knitting machine must receive an overlapped loop from at least one guide at every knitting cycle, otherwise it will press-off the fabric.

The underlap (U in Fig.34.2 and Fig.34.3) shog occurs across the side of the needles remote from the hooks on the front of single-needle bar, and in the center of double-needle bar, warp knitting machines. It supplies the warp yarn between one overlap and the next. The underlap shog generally ranges from 0 to 3 needle spaces, but it might be 14 needle spaces or more depending upon the design of the machine and the fabric structure (although efficiency and production speed will be correspondingly reduced with long underlaps).

Underlaps as well as overlaps are essential in warp knitted structures in order to join the wales of loops together but they may be contributed by different guide bars.

An open lap (Fig.34.2) is produced either when a subsequent underlap is in the same direction as the preceding overlap or an underlap is omitted so that the overlap of the next knitting cycle commences in the needle space where the previous overlap finished.

A closed lap (Fig. 34.3) is produced when a subsequent underlap shogs in the opposite direction to the preceding overlap, thus lapping the same yarn around the back as well as around the front of the needle. Closed laps are heavier, more compact, more opaque, and less extensible than open laps produced from the same yarn at a comparable knitting quality.

A course

Knitted loops are arranged in rows, roughly equivalent to the weft and warp of woven structures. These are termed "courses" and "wales" respectively.

A course is a predominantly horizontal row of needle loops (in an upright fabric as knitted) produced by adjacent needles during the same knitting cycle, as shown in Fig.34.4.

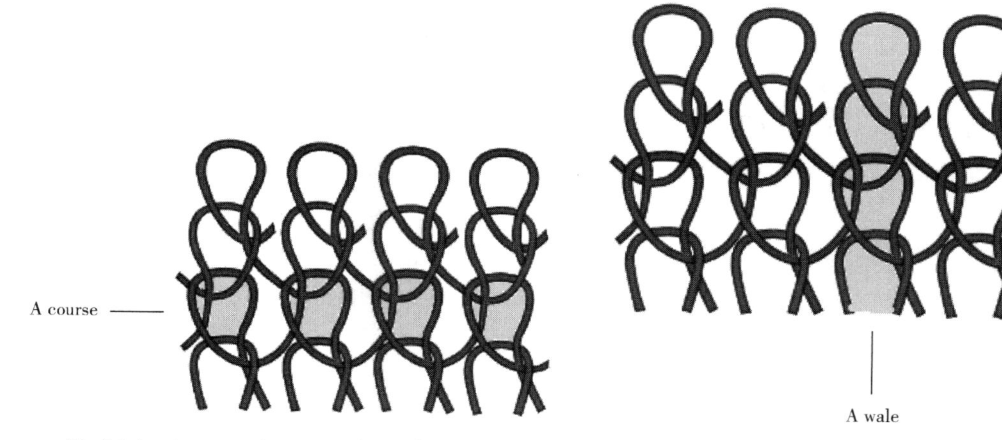

Fig.34.4 A course in a warp knitted structure

Fig.34.5 A wale in a warp knitted structure

A wale

A wale (Fig.34.5) is a predominantly vertical column of intermeshed needle loops generally produced by the same needle knitting at successive (not necessarily all) knitting cycles. A wale commences as soon as an empty needle starts to knit.

A pattern row

A pattern row is a horizontal row of needle loops produced by adjacent needles in one needle bed. In plain weft knitted fabric this is identical to a course but in more complex fabrics a pattern row may be composed of two or more course lengths. In warp knitting, every loop in a course is usually composed of a separate yarn.

Stitch density

Stitch density refers to the total number of loops in a measured area of fabric and not to the length of yarn in a loop (stitch length). It is the total number of needle loops in a given area (such as a square inch, or three square centimeters). The figure is obtained by counting the number of courses or pattern rows in one inch (or three centimeters) and the number of wales in one inch (or three centimeters), then multiplying the number of courses by the number of wales. (Using a measurement of three centimeters rather than one, is preferable for accuracy in counting).

Stitch density gives a more accurate measurement than does a linear measurement of only courses or only wales. Tension acting in one direction might produce a low reading for the courses and a high reading for the wales; when they are multiplied together this effect is cancelled out. Pattern rows rather than courses may be counted when they are composed of a constant number of courses.

The face loop stitch

Take the case of a weft knitted structure, the face side of the stitch (Fig.34.6) shows the new loop coming towards the viewer as it passes over and covers the head of the old loop. Face loop stitches tend to show the side limbs of the needle loops or overlaps as a series of interfitting "V"s.

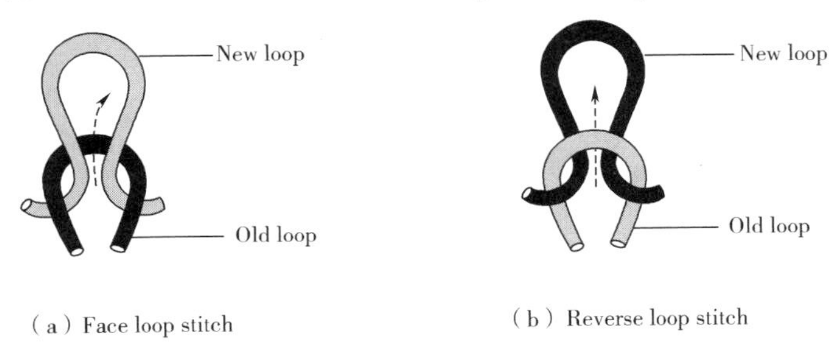

(a) Face loop stitch (b) Reverse loop stitch

Fig.34.6 Face loop stitch and reverse loop stitch

The reverse loop stitch

This is the opposite side of the stitch to the face loop-side and shows the new loop meshing away from the viewer as it passes under the head of the old loop, as Fig.34.6 (b) shows. Reverse stitches show the sinker loops in weft knitting and the underlaps in warp knitting most prominently on the surface.

Single-faced structures

Single-faced structures are produced in warp and weft knitting by the needles operating as a single set. Adjacent needles will thus have their hooks facing towards the same direction and the heads of the needles will always draw the new loops downwards through the old loops in the same direction.

The under-surface of the fabric on the needles (termed the technical face or right side) will thus only show the face stitches in the form of the side limbs of the loops or overlaps as a series of interfitting "V"s. The upper surface of the fabric on the needles (termed the technical back or left side) will show reverse stitches in the form of sinker loops or underlaps as well as the heads of the loops.

Double-faced structures

Double-faced structures are produced in weft and warp knitting when two sets of independently-controlled needles are employed with the hooks of one set knitting or facing in the opposite direction to the other set. The two sets of needles thus draw their loops from the same yarn in opposite directions, so that the fabric, formed in the gap between the two sets, shows the face loops of one set on one side and the face loops of the other set on the opposite side.

The two faces of the fabric are held together by the sinker loops or underlaps, which are inside the fabric so that the reverse stitches tend to be hidden. The two faces may be knitted from different yarns and the two fabrics thus formed may only occasionally be joined together. Sometimes the two faces are cohesively produced and are far enough apart for the connecting sinker loops or underlaps to be severed in order to produce two single-faced fabrics.

The knitted loop structure may not always be noticeable because of the effect of structural fineness, fabric distortion, additional pattern threads or the masking effect of finishing processes. The properties of a knitted structure are largely determined by the interdependence of each stitch to its neighbours on either side and above and below it.

New Words and Expressions

knitting 针织
intermesh 相互穿套
pin 棒针
loop 线圈
gauge 机号
weft knitting 纬编
warp knitting 经编
course 横列
wale 纵行
stitch 线圈
head 圈头，针编弧
legs or side limbs 圈柱
sinker loop 沉降弧
needle loop 圈干
overlap 针前垫纱，圈干
underlap 针背垫纱，延展线
open lap 开口线圈
closed lap 闭口线圈
warp guide 导纱针
knitting cycle 编织循环
pattern row 花型横列
stitch density 线圈密度
face loop stitch 正面线圈
reverse loop stitch 反面线圈
single faced structure 单面织物
double faced structure 双面织物

Notes to the Text

1. and knitted poor quality woolen stockings with a gauge of only 8 needles per inch: “gauge”是指针织机器的机号，其概念为每英寸安装的织针数量，单位为针/英寸。
2. The Englishman William Lee invented the first stocking hand frame in 1589: “frame”指“机器”，“hand frame”指“手摇横机”。
3. The knitted loop structure may not always be noticeable because of the effect of structural

fineness, fabric distortion, additional pattern threads or the masking effect of finishing processes: 由于针织结构比较细小，织物的变形、额外使用的花式纱线或者后整理过程的遮盖效应等，使得针织线圈结构并不是很容易被观察到。

Questions to the Text

1. What are the differences between weaving and knitting technology?
2. What is the basic unit in knitted fabric?
3. What are the components in weft knitted stitch and warp knitted stitch respectively?
4. What is the stitch density?

Lesson Thirty-Five

Knitting Machines and Principles *

The main features of the knitting machine

A knitting machine is an apparatus for applying mechanical movement, either hand or power derived, to primary knitting elements, in order to convert yarn into knitted loop structures. The machine incorporates and co-ordinates the action of a number of mechanisms and devices, each performing specific functions that contribute towards the efficiency of the knitting action.

The knitted machine types could be classified according to either the direction of yarn feeding or the amount of the needle bed, or the shape of needle bed, or the needle type applied on the machines, or the fabrics produced on the machines. Machines may range from high-production, limited-capability models to versatile, multi-purpose models having extensive patterning capabilities. The more complex the structure being knitted, the lower the knitting speed and efficiency. The simplest of the knitting machines would be hand-powered and manipulated whereas power driven machines may be fully automatically-programmed and controlled from a computer system.

The main features of a knitting machine (Fig.35.1) are as follows.

(1) The frame or carcass, normally free standing and either circular or rectilinear according to needle bed shape, provides the support for the majority of the machine's mechanisms.

(2) The machine control and drive system co-ordinates the power for the drive of the devices and mechanisms.

(3) The yarn supply consists of the yarn package or beam accommodation, tensioning devices, yarn feed control and yarn feed carriers or guides.

(4) The knitting system includes the knitting elements, their housing, drive and control, as well as associated pattern selection and garment-length control device (if equipped).

(5) The fabric take-away mechanism includes fabric tensioning, wind-up and accommodation devices.

(6) The quality control system includes stop motions, fault detectors, automatic oilers and lint removal systems.

The needle

The hooked metal needle is the principal knitting element of the knitting machine. There are three mechanical needles employed in modern knitting machines since the first frame was invented.

Fig.35.1 Face loop stitch and reverse loop stitch

They are bearded needle (spring needle), latch needle and compound needle.

The bearded or spring needle was the first type of needle to be produced. It is the cheapest and simplest type to manufacture as it is made from a single piece of metal, in machine gauges as fine as 60 needles per inch, with the needles being pliered to ensure accurate needle spacing.

The bearded needle is essentially a frame needle, the needles being fixed to move collectively with the straight needle bar or being attached to a circular frame and revolving with it.

There are six main parts of the bearded needle (Fig.35.2).

1—The stem, around which the needle loop is formed.

2—The head, where the stem is turned into a hook to draw the new loop through the old loop.

3—The beard, which is the curved downwards continuation of the hook that is used to separate the trapped new loop inside from the old loop as it slides off the needle beard.

4—The eye, or groove, cuts in the stem to receive the pointed tip of the beard when it is pressed, thus enclosing the new loop.

5—The shank, which may be bent for individual location in the machine or cast with others in a metal "lead".

6—The tip, which is the free end of the hook to touch the groove in the stem when the hook is closed.

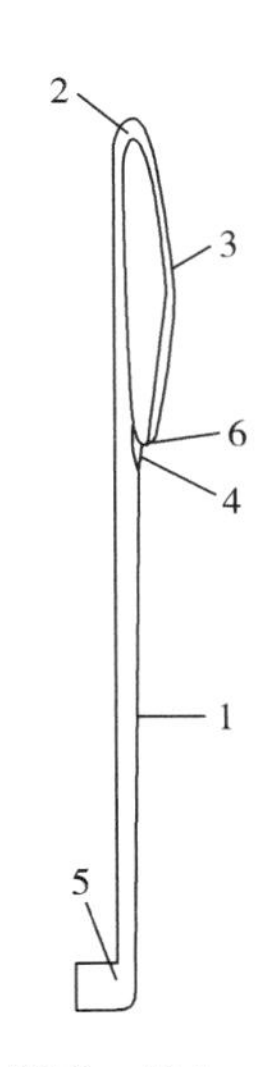

Fig.35.2 Main part of the bearded needle

The knitting action of the bearded needle is illustrated in Fig.35.3.

1—The needle is in the (so-called) rest position, with the previously formed loop (a) held on its stem and covered by the hook.

2—The loop is cleared from the needle hook to a lower position on the needle stem.

专业微课堂：成圈过程

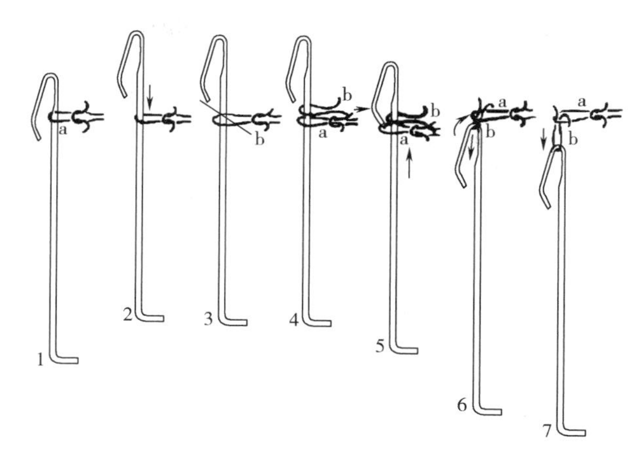

Fig.35.3 Basic knitting action of a bearded needle

3—The new yarn (b) is fed to the needle hook at a higher position on the needle stem than the position of the previous ("old") loop.

4—The yarn is formed into a "new" loop.

5—The hook is closed, enclosing the new loop and excluding and landing the old loop onto the outside of the closed hook.

6—The new loop (b) is drawn through the head of the old loop (a). Simultaneously the old loop slides off the closed hook of the needle and is cast-off or knocked-over.

7—The old loop now hangs from the feet of the fully formed new loop and the knitting cycle starts again.

The latch needle was more expensive and intricate than the bearded needle. It was more prone to making needle lines as it slided in its trick, particularly if the latch was damaged or there was dirt in the trick. It is now accepted that precision-manufactured latch needles can knit structures of the higher quality.

There are main five parts of latch needle, see Fig.35.4.

1—The hook, which draws and retains the new loop.

2—The lath, which is the closing element of the hook.

3—The rivet, which may be plain or threaded. This has been dispensed with on most plate metal needles, by pinching in the slot walls to retain the latch blade.

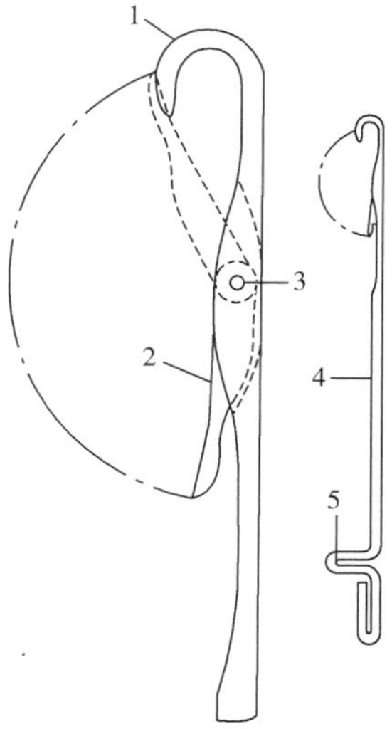

Fig.35.4 Main part of the latch needle

4—The stem, which carries the loop in the clearing or rest position.

5—The butt, which enables the needle to be reciprocated when contacted by cam profiles on either side of it, forming a track.

Fig.35.5 shows the positions of the latch needle passing from the lower part to the highest part of the cam track.

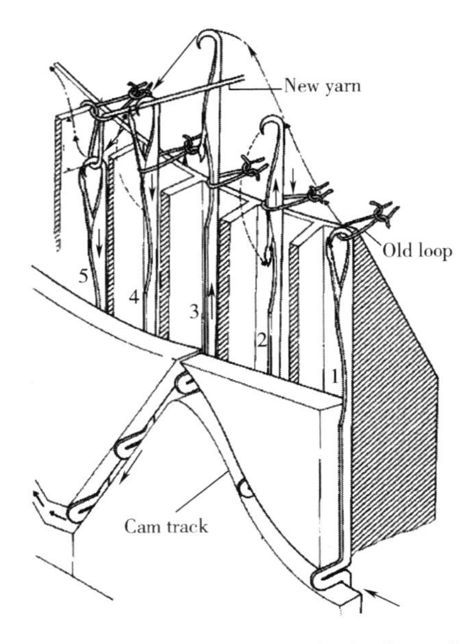

Fig.35.5 Knitting action of the latch needle

1—The rest position. The head of the needle hook is level with the top of the verge of the trick.

2—Latch opening. As the needle butt passes up the incline of the clearing cam, the old loop slides inside the hook and contacts the latch, turning and opening it.

3—Clearing height. When the needle reaches the top of the cam, the old loop is cleared from the hook and latch on to the stem.

4—Yarn feeding and latch closing. At this point the new yarn is fed through a hole in the feeder guide to the descending needle hook, the old loop contacts the underside of the latch, causing it to close on to the hook.

5—Knocking-over and loop length formation. As the head of the needle descends below the top of the trick, the old loop slides off the needle and the new loop is drawn through it.

Compound needles (Fig. 35.6) consist of two separately-controlled parts—1-the open hook and 2-the sliding closing element (tongue, latch, piston, plunger).

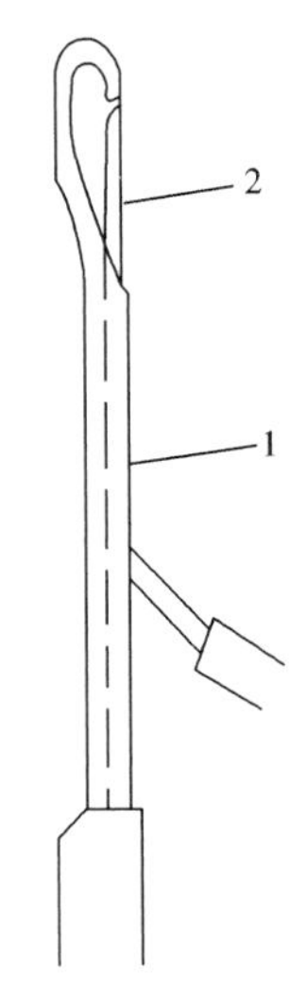

Fig.35.6 Main part of the compound needle

The two parts rise and fall as a single unit but, at the top of the rise, the hook moves faster to open the hook and at the start of the fall the hook descends faster to close the hook. It is easier to drive the hooks and tongues collectively from two separate bars in warp knitting than to move each hook and tongue individually, as in weft knitting.

Two types of compound needle have been employed in warp knitting machines. The tubular pipe needle has its tongue sliding inside the tube of the open hook. Development ceased after 1950s and bearded needle tricot machines recaptured their market with higher speeds, only to be later outpaced by a more efficient type of compound needle, the slide compound needle.

The open-stem "pusher type" or slide needle has a closing wire or tongue that slides externally along a groove on the edge of the flat hook member. This needle is now preferred because it is simpler, cheaper, more compact and each of the two parts can be separately replaced.

New Words and Expressions

yarn supply 供纱
knitting system 编织系统
fabric take-away mechanism 织物牵拉机构
bearded needle 钩针
latch needle 舌针
compound needle 复合针
hook 针钩
stem 针杆
butt 针踵
latch 针舌
cam track 三角轨道
rest position 起始位置
clearing 退圈
knocking-over 脱圈

Notes to the Text

The frame or carcass, normally free standing and either circular or rectilinear according to needle bed shape：这里指针织机根据针床的形状，可以分为圆形针织机和平型针织机。

Questions to the Text

1. What are the main features of the knitting machine?
2. How many kinds of needles employed on knitting machines? What are they?
3. What are the main parts of the latch needle?

Lesson Thirty-Six

Basic Knitted Structures*

The basic weft knitted structures

Plain, rib, interlock and purl are the base structures to form almost all weft knitted fabrics and garments. Each is composed of a different combination of face and reverses meshed stitches, knitted on a particular arrangement of needle beds. Each primary structure may exist alone, in a modified form with stitches other than normal cleared loops, or in combination with another primary structure in a garment-length sequence.

Plain is produced by the needles knitting as a single set, drawing the loops away from the technical back and towards the technical face side of the fabric. Plain can be unroved from the course knitted last by pulling the needle loops through from the technical back, or from the course knitted first by pulling the sinker loops through from the technical face side. Loops can be prevented from unroving by binding-off.

Rib requires two sets of needles operating in between each other so that wales of face stitches and wales of reverse stitches are knitted on each side of the fabric. The simplest rib fabric is 1×1 rib. Rib has a vertical cord appearance because the face loop wales tend to move over and in front of the reverse loop wales. As the face loops show a reverse loop intermeshing on the other side, 1×1 rib has the appearance of the technical face of plain fabric on both sides until stretched to reveal the reverse loop wales in between.

Interlock was originally derived from rib but requires a special arrangement of needles knitting back-to-back in an alternate sequence of two sets, so that the two courses of loops show wales of face loops on each side of the fabric exactly in line with each other, thus hiding the appearance of the reverse loops. Interlock has the technical face of plain fabric on both sides, but its smooth surface cannot be stretched out to reveal the reverse meshed loop wales because the wales on each side are exactly opposite to each other and are locked together.

Purl is the only structure having certain wales containing both face and reverse meshed loops. A garment-length sequence, such as a ribbed half-hose, is defined as purl, whereas smaller sections of its length may consist of plain and rib sections. Purl structures have one or more wales which contain both face and reverse loops. This can be achieved with double-ended latch needles or by rib loop transfer from one bed to the other, combined with needle bed racking.

专业微课堂：
针织面料

The basic warp knitted structures

Pillar stitch, tricot lapping, atlas lapping and double needle overlaps structure are the base structures to form almost all warp knitted fabrics and products.

In the pillar or chain stitch, the same guide always overlaps the same needle. This lapping movement will produce chains of loops in unconnected wales, which must be connected together by the underlaps of a second guide bar.

Generally, pillar stitches are made by front guide bars, either to produce vertical stripe effects or to hold the inlays of other guide bars into the structure. Open-lap pillar stitches are commonly used in warp knitting. They can be unroved from the end knitted last. Closed-lap pillar stitches are employed on crochet machines because the lapping movement is simple to achieve and is necessary when using self-closing carbine needles, which must always be fed with yarn from the same side.

Tricot lapping is based on the balanced advance and return lapping in two courses. Tricot lapping or 1×1 is the simplest of these movements, producing overlaps in alternate wales at alternate courses with only one thread crossing between adjacent wales. Two threads will cross between wales with a 2×1 or cord lap, three threads with a 3×1 or satin lap, four threads with a 4×1 or velvet lap, and so on.

Atlas lapping is a movement where the guide bar laps progressively in the same direction for a minimum of two consecutive courses, normally followed by an identical lapping movement in the opposite direction. Usually, the progressive lapping is in the form of open laps and the change of direction course is in the form of a closed lap, but these roles may be reversed. From the change of direction course, tension tends to cause the heads of the loops to incline in the opposite direction to that of the previous lapping progression. The change of direction course is normally tighter and the return progression courses cause reflected light to produce a faint, transverse shadow, stripe effect.

In double needle overlaps structure, the stitches are formed simultaneously by the same warp end on two adjacent needles in one course. The tension of the yarn is higher than that in other warp knitted structures. The underlap between the double overlaps has the appearance of a sinker loop. So the properties of the fabric are just between those of normal warp knitted and weft knitted fabrics.

New Words and Expressions

plain　纬平针	pillar stitch　编链
rib　罗纹	tricot lapping　经平
interlock　双罗纹	atlas lapping　经缎
purl　双反面	double needle overlaps structure　重经

Notes to the Text

1. The simplest rib fabric is 1×1 rib：最简单的罗纹组织是 1×1 罗纹，是指由正面线圈纵行及反面线圈纵行 1 隔 1 间隔编织而成的组织。
2. Tricot lapping or 1×1 is the simplest of these movements：1×1 经平是最简单的经平类组织，是指针前垫纱和针背垫纱均为 1 针距。

Questions to the Text

1. What are the four basic weft knitted structures?
2. What are the four basic warp knitted structures?
3. What is the simplest rib structure?

Lesson Thirty-Seven

Fabric Finishing and Dyeing

Finishing

Consumers would not readily accept most cloths in the state in which they emerge from the weaving, knitting or other manufacturing processes. Unfinished fabrics, referred to as greige or grey goods, contain many impurities, have a harsh hand and little esthetic appeal. Before reaching the consumer market, cloth must undergo at least one and usually several finishing processes. The purpose of these finishes is to impart specific properties to the cloth to make it more appealing to the consumer. The end result may be improved esthetics, better performance, increased durability, resistance to insects, molds, and fungi, improved safety or protection for the user, the addition of color and design, or simply the preparation of the cloth for further finishing processes.

Finishes may be applied to cloth by mechanical means. Although many of these techniques have been in use for thousands of years, improvements in machinery design, advances in chemical technology, and the discovery of new types of finishes have greatly improved the performance of textile products while reducing the cost to the consumer. The major advances have been in the field of chemical finishes. Since 1950, developments in finishing technology such as durable-press polyester/cotton blends, shrink resistant wool, and flame-retardant treatments have decreased maintenance requirements, improved performance, and enhanced the safety of textile materials. Significant new chemical finishes are likely to be developed in the next decade. The finishing processes performed on cloth from the point of manufacture until it is ready for sale may be divided into three broad categories: general finishes, functional finishes, and the application of color.

The two examples of functional finishing are given as follows.

(1) Flame retardant finishes[Fig.37.1(a)]: A flame retardant textile may or may not ignite after exposure to a heat source. However, if it does ignite, it will burn or smolder for only a short time after the heat source is removed and will self-extinguish. Three general methods are available for producing cloth with improved flammability: ① flame retardant chemical finishes may be applied to the surface of cloth, ② they may be added to the polymer at some step in the synthesis process prior to extrusion of the polymer into fibers, or ③ inherently flame resistant fibers may be developed and converted into cloth. Most textile flame retardants change or interrupt the normal thermal decomposition process of the polymer. They may function in the solid substrate (cloth) to lower the decomposition temperature of the fiber and change the nature and amount of volatile combustibles formed, or they may function in the vapor phase. For the flame retardant chemical to

be effective, it must be available at the point at which the polymer decomposes.

(2) Water-repellent finishes [Fig.37.1(b)]: A water-repellent cloth is one in which the yarns are coated with a hydrophobic (water-hating) compound, but the pores between the warp and filling yarns are not filled. The cloth is permeable to water vapor and air. The ability of a surface to resist wetting is a function of the chemical nature of the surface, the surface roughness, the porosity of the surface, and the presence of other water molecules on the surface. The major problem of water-repellent finishes is the durability to laundering and dry-cleaning. Current washing and dry-cleaning practices do not completely remove detergents from cloth, so the retention of durable water-repellent properties in the presence of detergents is difficult. Initially, the water-repellent properties of raincoats are very good but after several renovation procedures, they no longer resist wetting and must be retreated.

(a) A flame retardent textile

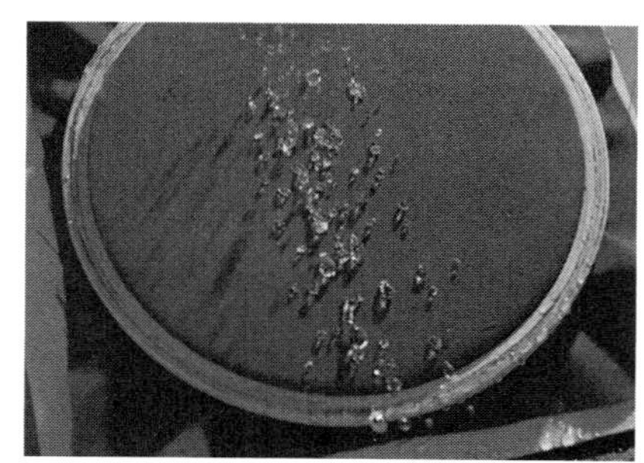

(b) A water-repellent textile

Fig.37.1 Flame retardant and water-repellent textiles

Dyeing

It would be quite true to say that the color of a textile fabric is the first property which is noticed and frequently is the first factor governing fabric choice. Any person selecting a fabric or a garment will either have a definite color in mind or will be initially attracted by a color and will investigate from the point of attraction. Color is applied to textile materials in the form of dyestuffs which can be used in dyeing process. In its simplest form consists of the immersion of the textile fabric material in a solution of the dyestuff in water. When the temperature is raised sufficiently the dyestuff passes from the solution into the textile material and colors it uniformly.

The dyestuffs used on textiles were all obtained from synthetic and natural sources. Synthetic dyes provided a wider, clearer and brighter range of colors. They could be made pure and constant in color value. Natural dyes varied greatly in quality and lack of precise chemical knowledge in applying them made it difficult to produce constant matching color tones, they also contained impurities which hindered operations, and once the chemists could isolate and prepare the pure coloring, preparation became simpler and quicker; instead of days to prepare a vat, it could be done in hours. Synthetic dyes were also cheaper than natural dyes particularly in terms of time and trouble.

Dyeing is complicated for two main reasons.

(1) It is not sufficient merely to apply color to textile fabric; it must stay in the fabric and not fade quickly when exposed to light; it must not bleed or run out of the fabric if it is washed; it must not rub off onto the wearer. These are some of the main considerations but they can be summed up by saying that color fastness to various agencies is desirable.

(2) The various natural and man-made fibers vary in their affinity, or capacity to take dyestuffs.

Dyeing is not just a method of staining a textile fiber with a color solution; it can be a highly complex operation involving strict control of time, temperature and chemical processes but the ultimate object is to color the textile material to the best degree of color and fastness that the fiber or fibers, the selected dyestuff and process of application will allow, taking cost into consideration.

New Words and Expressions

greige　坯布，坯绸
grey goods　坯布
harsh hand　粗糙的手感
esthetic appeal　美的外观
polyester/cotton blend　涤/棉混纺
flame retardant　阻燃，阻燃剂
general finish　一般整理
functional finish　功能整理
ignite　点燃，使燃烧
flame retardant textile　阻燃纺织品
porosity　多孔性
color value　色值
bleed　渗色
affinity　亲和性

Notes to the Text

1. cloth in the state in which it emerges from the weaving, knitting or other manufacturing processes：指未经整理，直接从织机或其他设备上制成的下机坯布。
2. They may be white, grey or colored if dyed yarns or fibers have been used in their manufacture：坯布可以是白色的、本色的，如果经过了纱线染色或纤维染色，则可以是带有色彩的。"grey"可以代表灰色，但在本句中意思是纤维本来的颜色。
3. to make it more appealing to the consumer：使之对顾客更具吸引力。
4. For the flame retardant chemical to be effective, it must be available at the point at which the polymer decomposes：句中的"for the flame retardant chemical to be effective"可译为"为了使阻燃剂产生效果"，句中"available"可译为"发挥作用"。
5. color fastness to various agencies is desirable：任何领域都希望颜色具有较好的坚牢度。
6. the ultimate object is to color the textile material to the best degree of color and fastness that the fiber or fibers, the selected dyestuff and process of application will allow, taking cost into consideration：染色的最终目的是给纺织材料着色，色彩要好，牢度要高；这些是以纤维、染料以及染料的施加方式为基础的，还要考虑到成本。

Questions to the Text

1. Tell the types of finishes you know.

2. Why the grey goods must be finished?
3. What are the general methods for producing flame retardant fabrics?
4. Why washing will affect the effectiveness of water-repellency properties?
5. What are the differences between natural and synthetic dyes?
6. Why dyeing is a complicated process?

Lesson Thirty-Eight

Fabric Permanent Set and Shrinkage Control *

Fabric permanent set

The Commonwealth Scientific and Industrial Research Organization in Australia has developed a process for permanent setting of wool in creases and pleats. The process consists of impregnating the area to be creased with a solution of a chemical compound. When the crease is pressed under controlled conditions, the chemical compound inside the wool fibers locks them in their creased position and prevents them from exercising their normal recovery powers. Another IWS permanent set process can be applied to a trouser crease from a hand applicator or a machine resembling a sewing machine. After pressing and finishing in the normal way the garment is turned inside out to expose the inside of the crease and a thin layer of resin is applied in the groove of the crease. When pressed again the resin is cured, leaving a crease which is fast to washing and dry cleaning.

Cotton and linen are not appreciably more stretchable when hot and moist than when dry, but viscose rayon, silk and wool can be stretched considerably if required to. If a fabric is dried whilst being held in a stretched state it will be temporarily set in that state, but if later it is treated at or near the temperature at which it was set it will lose its temporary set and will begin to relax in an effort to reach normal equilibrium. This relaxation can occur in pressing, washing, drying or even in wear conditions and the most common result observed is shrinkage, or distortion of garment shape caused by uneven shrinkage. Obviously there will be a greater likelihood of shrinkage in this way in stretchable fabrics and for this reason the finisher must take care when processing stretchable fabrics made from wool, silk and rayon not to set them in a greatly stretched state. The clothing manufacturer should also test every piece of knitted fabric since the relaxation shrinkage may vary even within a finishing batch of apparently identical fabric.

Synthetic fibers such as nylon and polyester can be set by heat and moisture, or by dry heat, in a permanent manner, i. e. , unless the fabric is again subjected to the setting temperature and conditions. Modern setting is done at temperature well above boiling point, which of course demands strict process control, to avoid the risk of accidental release of set. As a further safeguard washing and drying temperatures for synthetic fiber fabrics must be kept well below boiling. The main reason for this is that at or near boiling temperature unwanted crease can be accidentally set in these fabrics. The main set would not be disturbed but the appearance of a fabric can be completely spoilt by the imposition of creases which resist all attempts to remove by ironing.

Fabric shrinkage control

The shrinkage of fabrics on washing is a problem which existed long before man-made fibers and certain traditional attitudes to textile materials are a direct consequence of the shrinkage problem. The cheapness and versatility of cotton has meant that the fiber has for a very long time been used in a wide and varied range of fabrics which needed regular washing and the shrinkage which inevitably followed was a difficult problem. Many ideas were tried to overcome it but without a great deal of success until the principles of controlled compressive shrinkage treatments were discovered.

Controlled compressive shrinkage treatments work by compressing fabrics the amount they actually shrink so that the yarns are in their shrunk position before the fabric is used, and this can be done without spoiling the finished appearance of the fabric. Different types of fabrics have differing shrinkage potential so that testing of fabric for shrinkage is an essential part of treatment. When the series of tests reveals the shrinkage potential of the fabrics, the processing equipment is then set to compress the fabric to zero shrinkage, with a deliberate margin of plus or minus 1% which prevents any risk of over-treatment.

It was seen that because cotton is not very stretchable it has no great relaxation potential. Wool is quite different; it has considerable powers of extension and recovery so that tensions are accumulated during manufacture and finishing. When all processes are completed the wool fibers if left alone will try to creep back to their unstressed length. They may be prevented from doing this by temporary set, but if and when the set is released by heat and moisture contraction will begin. Because of the stretching capacity of wool, and because relaxation shrinkage is completely independent of washing it presents a great problem in wool fabric use.

As far as woven fabrics are concerned deliberate relaxation processes such as London shrinking are the means of either removing this form of shrinkage or reducing it to negligible proportions. The principle of this process is moistening of the fabric and then drying without tension. As this can be rather slow there are quite a few speeded up versions involving the use of heat and steam. Relaxation processes are not a normal routine of finishing so that it is unsafe to assume that a fabric has been relaxed. Many clothing manufacturers have installed shrinkage machines and treat all their wool fabrics themselves irrespective of whether relaxation was part of fabric treatment or not. This is to avoid shrinkage and distortion occurring during making-up when pressing operations are used.

专业微课堂：
织物整理

Knitted wool fabrics are also subject to relaxation shrinkage, but it is an area shrinkage and not as directional as in woven fabric. Because of the greater elasticity of the knitted structure control of dimensional stability of knitted fabrics is more difficult. Great attention must be paid to the correct balance of the fabric structure as it leaves the machine so that the structure is correctly

思政微课堂：
香云纱

compacted and if this is done the finisher can produce a relaxed fabric.

New Words and Expressions

International Wool Secretariat(IWS)　国际羊毛咨询处
crease　褶裥
relaxation shrinkage　松弛回缩
making-up　裁剪，缝制

Notes to the Text

1. Commonwealth Scientific and Industrial Research Organization in Australia：澳大利亚联邦科学与工业研究组织。
2. clothing manufacturer：服装生产商。
3. boiling point：指水的沸点。
4. The main reason for this is that at or near boiling temperature unwanted crease can be accidentally set in these fabrics. The main set would not be disturbed but the appearance of a fabric can be："the main set"指的是所需要的定形。
5. and certain traditional attitudes to textile materials：以及一些对纺织材料的传统处理方式。"attitude"的原意是"态度"，在这里译为"处理方式"。
6. Controlled compressive shrinkage treatments work by compressing fabrics："work"的意思是"产生效果，起作用"。
7. Because of the stretching capacity of wool, and because relaxation shrinkage is completely independent of washing it presents a great problem in wool fabric use："it presents a great problem in wool fabric use"是句子的主体。
8. London shrinking：伦敦预缩法(织物润湿，在无张力下自然干燥)。
9. area shrinkage：面积收缩，双向收缩。

Questions to the Text

1. Describe the permanent setting processes given by CSIRO in Australia and IWS.
2. Why care must be taken for the fabrics being held in a stretchable state?
3. How to control the compressive process in order to get zero shrinkage?
4. What are the differences between cotton and wool fabric in shrinkage?
5. What is London shrinkage?
6. What is the relaxation shrinkage of the knitted wool fabrics?

Lesson Thirty-Nine

语言微课堂

Nonwovens

The term "nonwoven" originated in the late 1930s when nonwovens were often regarded as low-price substitutes for traditional textiles and were generally made from dry-laid carded webs using converted textile processing machinery. The yarn spinning stage is omitted in the nonwoven processing of staple fibers, while bonding (consolidation) of the web by various methods, chemical, mechanical or thermal, replaces the weaving (or knitting) of yarns in traditional textiles. As markets were developed, and producers became adept at matching properties to end uses, nonwovens were able to replace woven materials and yet be cheap enough to be discarded after single or limited use, the primary focus then was sanitary and medical products. Therefore, the relatively young nonwoven industry has established itself as a distinct and viable entity.

EDANA (the European Disposables and Nonwovens Association) defines a nonwoven as "a manufactured sheet, web or batt of directionally or randomly orientated fibers, bonded by friction, and/or cohesion and/or adhesion", but goes on to exclude a number of materials from the definition, including paper, products which are woven, knitted, tufted or stitchbonded (incorporating binding yarns or filaments), or felted by wet-milling, whether or not additionally needled. INDA, North America's Association of the Nonwoven Fabrics Industry, describes nonwoven fabrics as "sheet or web structures bonded together by entangling fibers or filaments, by various mechanical, thermal and/or chemical processes". These are made directly from separate fibers or from molten plastic or plastic film.

Dry, wet and polymer-laid nonwovens

Generally, according to the different production methods, nonwovens can be divided into three major areas: dry-laid, wet-laid and polymer-laid (encompassing the spunmelt technologies of spunbond, meltblown, flashspun, fibrillated film and electrospinning). Fig.39.1 is the overview of nonwoven materials.

In dry-laid web formation, fibers are carded (including carding and cross-lapping) or aerodynamically formed (air-laid) and then bonded by mechanical, chemical or thermal methods. These methods are mechanical bonding (including needlepunching, hydroentanglement and stitchbonding), thermal bonding (including ultrasonic bonding, through air bonding and calender bonding) and chemical bonding (including saturation bonding, spray bonding, print bonding, foam bonding and solvent bonding).

Paper-like nonwoven fabrics are manufactured with machinery designed to manipulate short

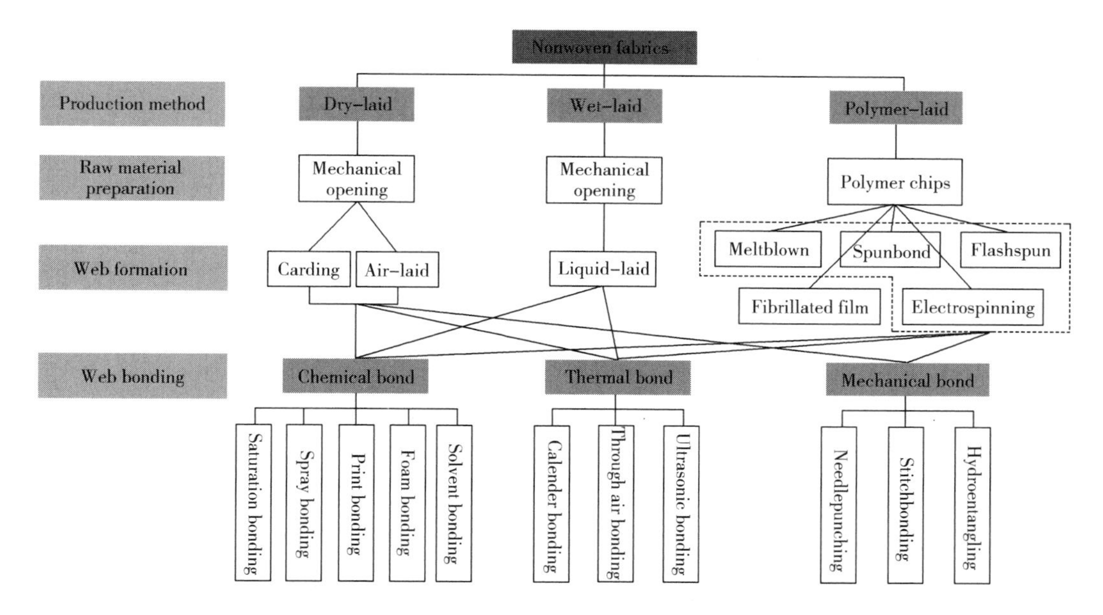

Fig.39.1 Overview of nonwoven materials

fibers suspended in liquid and are referred to as "wet-laid". To distinguish wet-laid nonwovens from wet-laid papers, a material is regarded by EDANA as a nonwoven if more than 50% by mass of its fibrous content is made up of fibers (excluding chemically-digested vegetable fibers) with a length to diameter ratio greater than 300, or more than 30% fiber content for materials with a density less than 0.40g/cm^3. This definition excludes most wet-laid glass fiber constructions which sectors of the industry would class as nonwovens. The use of the wet-laid process is confined to a small number of companies, being extremely capital intensive and utilizing substantial volumes of water. In addition to cellulose papers, technical papers composed of high performance fibers such as aramids, glass and ceramics are produced.

专业微课堂：
非织造材料成形方式

思政微课堂：
非织造材料的
发展和现状

Polymer-laid or "spunmelt" nonwoven fabrics are produced by extrusion spinning processes, in which filaments are directly collected to form a web instead of being formed into tows or yarns as in conventional spinning. As these processes eliminate intermediate steps, they provide opportunities for increasing production and cost reductions. In fact, melt spinning is one of the most cost efficient methods of producing fabrics. Commercially, the two main polymer-laid processes are spunbonding (spunbond) and meltblowing (meltblown).

Spunbond fabrics are produced by depositing extruded, spun filaments onto a collecting belt in a uniform random manner followed by bonding the fibers. Guiding by the developing trend of fine and composite, some novel spunbond technologies become the hot spots, especially bicomponent spunbond nonwovens, which possess the incomparable advantages on cost-saving, function and product performance. Bicomponent spunbonded andspunlaced technology is a way to produce superfine fiber nonwovens.

Meltblowing is a process for producing fibrous webs or articles directly from polymers or resins using high-velocity hot air or another appropriate force to attenuate the filaments. This process is unique because it is used almost exclusively to produce microfibers rather than the size of normal textile fibers.

Web formation

In all nonwoven web formation processes, fibers or filaments are either deposited onto a forming surface to form a web or are condensed into a web and fed to a conveyor surface. The conditions at this stage can be dry, wet, or molten—dry-laid, wet-laid or polymer-laid (also referred to as spun-laid and spunmelt processes). Web formation involves converting staple fibers or filaments into a two-dimensional (web) or a three-dimensional web assembly (batt), which is the precursor for the final fabric. Their structure and composition strongly influence the dimensions, structures and properties of the final fabric. The fiber (or film) orientation in the web is controlled during the process using machinery adapted from the textile, paper or polymer extrusion industries.

The arrangement of fibers in the web, specifically the fiber orientation, governs the isotropy of fabric properties and most nonwovens areanisotropic.

Web bonding

Nonwoven bonding processes can be mechanical, chemical or thermal. The degree of bonding is a primary factor in determining fabric mechanical properties (particularly strength), porosity, flexibility, softness, and density (loft, thickness). Bonding may be carried out as a separate and distinct operation, but is generally carried out in line with web formation. In some fabric constructions, more than one bonding process is used. Mechanical consolidation methods include needlepunching, stitchbonding, and hydroentangling. The latter process has grown considerably in popularity over the past few years. Chemical bonding methods involve applying adhesive binders to webs by saturating, spraying, printing, or foaming techniques. Solvent bonding involves softening or partially solvating fiber surfaces with an appropriate chemical to provide self-or autogeneously bonded fibers at the cross-over points. Thermal bonding involves the use of heat and often pressure to soften and then fuse or weld fibers together without inducing melting.

New Words and Expressions

substitute 替代品
consolidation 固结;加固
adept 能手,内行
discard 丢弃
sanitary 卫生的
viable 可行的
entity 实体
batt 棉絮;絮垫

tuft 簇绒
stitchbond 缝编
felt 毡制品；把……制成毡
wet-milling 湿法缩绒
dry-laid 干法成网
polymer-laid 聚合物直接成网
spunbond 纺粘
meltblown 熔喷
flashspun 闪蒸
fibrillated film 膜裂
electrospinning 静电纺丝
carding 梳理
cross-lapping 交叉铺网
aerodynamically 空气动力学地
air-laid 气流成网
needlepunching 针刺
hydroentanglement 水刺
stitchbonding 缝编
ultrasonic bonding 超声波黏合
air bonding 热风黏合
calender bonding 热轧黏合
saturation bonding 饱和浸渍黏合
spray bonding 喷洒黏合
print bonding 印花黏合
foam bonding 泡沫黏合
wet-laid 湿法成网
aramid 芳香族聚酰胺
ceramic 陶瓷，陶瓷制品
tows 丝束
bicomponent 双组分的
spunlace 水刺
conveyor 输送机；传送带
spun-laid 纺丝成网
isotropy 各向同性
anisotropic 各向异性的
fuse 熔接
weld 焊接

Notes to the Text

1. nonwovens were often regarded as low-price substitutes for traditional textiles and were generally made from dry-laid carded webs using converted textile processing machinery：非织造布通常被视为传统纺织品的低价替代品，通常是使用经改造的纺织品加工机械由干式梳理纤维网制成的。
2. producers became adept at matching properties to end uses, nonwovens were able to replace woven materials and yet be cheap enough to be discarded after single or limited use, the primary focus then was sanitary and medical products：生产商开始熟练地将性能与最终用途相匹配，非织造布能够取代机织材料，而且价格足够便宜，可以在一次或有限的使用后丢弃，当时的主要焦点是卫生和医疗产品。
3. EDANA (the European Disposables and Nonwovens Association)：欧洲耗材及非织造布协会。
4. a manufactured sheet, web orbatt of directionally or randomly orientated fibers, bonded by friction, and/or cohesion and/or adhesion', but goes on to exclude a number of materials from the definition, including paper, products which are woven, knitted, tufted or stitchbonded (incorporating binding yarns or filaments), or felted by wet-milling, whether or not additionally needled：定向或随机排列的纤维通过摩擦、抱合、黏合或者这些方法的组合而相互结合制成的片状物、纤网或絮垫，不包括纸、机织物、针织物、簇绒织

物、带有缝编纱线的缝编织物以及湿法缩绒的毡制品(无论是否另外进行针刺)。

5. INDA(International Nowoven Disposables Association):国际非织造布和即用弃制品协会。
6. North America's Association of the Nonwoven Fabrics Industry:北美非织造布工业协会。
7. capital intensive:资本密集。

Questions to the Text

1. How many methods are being used to make nonwovens?
2. How many methods are being used for web formation?
3. How many methods are being used for web bonding?
4. What is the difference between nonwovens and paper?

Lesson Forty

Nonwovens for Personal Protective Equipment (PPE) *

The concept of personal protective equipment (PPE) for the surgeon has been in place for greater than 100 years. The aim has been to minimize the transmission of various pathogens and protect both the surgeon and the patient. Before Coronavirus Disease 2019 (COVID-19), the acronym PPE was not known by most people. Healthcare workers, first responders and industrial workers are familiar with PPE because they use it every day. There are several different types of PPE, including face masks, gowns, protective clothing, goggles, gloves, face shields, nonpowered filtering facepiece respirators (FFRs), and powered air-purifying respirators. Some of them made of nonwovens will be discussed in detail below.

The use of surgical masks during the COVID-19 pandemic has been a subject of debate. Surgical masks are popularly worn by the general public all year round in East Asian countries like China, Japan and South Korea, especially during allergy and flu seasons, to reduce the chance of spreading airborne diseases to others, and to prevent the breathing in of airborne pollens or dust particles created by air pollution (dust masks are more effective against pollution).

Thanks for the mask support of general public and effective measures, the pandemic was effectively controlled and all sectiors in China are striving to resume work and production in an orderly manner to minimize economic and social fallout from the COVID-19 outbreak. Nowadays, China has become the largest supplier of anti-epidemic materials in the world. China exported 224.2 billion masks between March and the end of December in 2020 to assist the international community in fighting COVID-19, official data showed on Jan 14, 2021. Among the exported masks, a total of 65 billion are for medical use, according to the Chinese General Administration of Customs (GAC). The country has exported 773 million medical protective suits and 2.92 billion pairs of surgical gloves during the same period to protect disease control personnel in the global fight against COVID-19. There is no doubt that China has made great contributions to fighting against the pandemic in the world.

During this pandemic, healthcare workers and others have been scrambling for two sought-after products in the PPE pandemic world: surgical face masks and the N95 respirator.

The surgical, medical, or procedure mask (face masks)

A surgical mask, also known as a medical face mask, were originally designed to protect medical personnel from accidentally breathing or swallowing in splashes or sprays of bodily fluids.

The mask is a PPE worn by health professionals during medical procedures to prevent airborne transmission of infections between patients and/or treating personnel, by blocking the transmission of pathogens (primarily bacteria and viruses) shed in respiratory droplets and aerosols from the wearer's mouth and nose. It acts as an additional barrier for the airway and is not usually designed (unless N95-rated) to completely prevent the wearer from inhaling smaller airborne pathogens, but could be still protective by filtering out and trapping most of the droplets that carry them.

Surgical face masks are made with nonwoven fabric, which has better bacteria filtration and air permeability while remaining less slippery than woven cloth. The material most commonly used to make them is polypropylene, either 20 or 25 grams per square meter in density. Masks can also be made of polystyrene, polycarbonate, polyethylene, or polyester. 20 grams per square meter mask material is made in a spunbond process, which involves extruding the melted plastic onto a conveyor. The material is extruded in a web, in which strands bond with each other as they cool. 25 grams per square meter fabric is made through meltblown technology, which is a similar process where plastic is extruded through a die with hundreds of small nozzles and blown by hot air to become tiny fibers, again cooling and binding on a conveyor. These fibers are less than a micron in diameter, with an electrostatic charge; that is, the fibers are electrets. The electret filter increases the chances that smaller particles will veer and hit a fiber, rather than going straight through (electrostatic capture).

Surgical masks are made up of a multi-layered structure (most 3-ply or 4-ply), generally by covering a layer of textile with nonwoven bonded fabric on both sides, as shown in Fig. 40.1. Nonwovens, which are cheaper to make and cleaner thanks to their disposable nature, are made with three or four layers. These disposable masks are often made with two filter layers effective at filtering out particles such as bacteria above 1 micron. The filtration level of a mask, however, depends on the fiber, the way it's manufactured, the web's structure, and the fiber's cross-sectional shape. Masks are made on a machine line that assembles the nonwovens from bobbins, ultrasonically welds the layers together, and stamps the masks with nose strips, ear loops, and other pieces. Completed masks are then sterilized before being sent out of the factory.

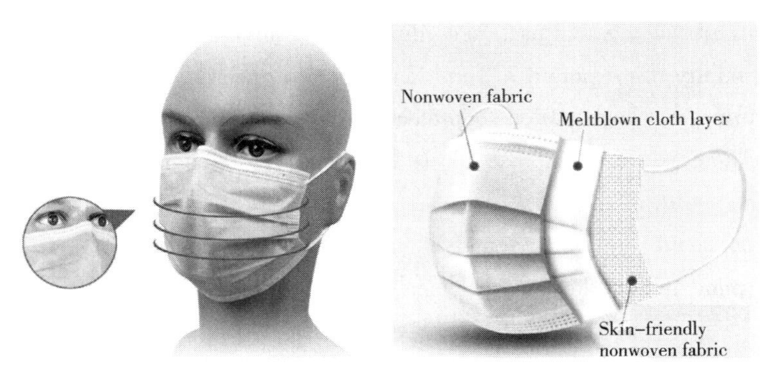

思政微课堂：
2020年中国出口
口罩2242亿只

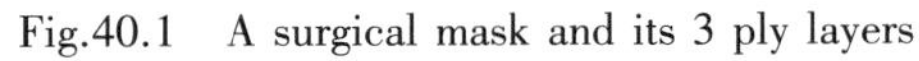
Fig.40.1 A surgical mask and its 3 ply layers

Surgical masks may be labeled as surgical, isolation, dental or medical procedure masks. Chinese health officials distinguish between medical (non-surgical) and surgical masks. Commonly seen surgical masks have different colour. Generally, the green blue (or pink) side of the mask (the fluid-repellant layer) is to be worn outward, with the white (absorbent) layer on the inside. With respect to some infections like influenza surgical masks could be as effective (or ineffective) as respirators, such as N95, KN95 or FFP masks; though the latter provide better protection in laboratory experiments due to their material, shape and tight seal.

Surgical masks for use in the US and the European Economic Area (EEA) conform to ASTM F2100 and EN 14683 respectively. In both standards, a mask must have a bacterial filtration efficiency (BFE) of more than 95%, simulated with particles of size 3.0μm. In China, two types of masks are common: surgical masks that conform to YY 0469 standard (BFE≥95%, PFE≥30%, splash resistance) and single-use medical masks that conform to YY/T 0969 standard (BFE ≥ 95%). Daily protective masks conforming to GB/T 32610 standard is yet another type of masks that can have similar appearance to surgical masks. These standards in China, USA & Europe for surgical masks are listed as below in Table 40.1.

Table 40.1 EN, ASTM & GB standards for medical face masks

Item	Europe(EN 14683)	US(ASTM F2100)	China(YY 0469)	China(YY 0969)
Filtration(BFE)	≥95%(Type Ⅰ) ≥98%(Type Ⅱ, ⅡR)	≥95%(Level 1) ≥98%(Level 2, 3)	≥95%(ASTM F2101)	≥95% (ASTM F2101)
Filtration(PFE)	N/A	≥95%(Level 1) ≥98%(Level 2, 3)	N/A	N/A
Pressure drop/(Pa/cm^2)	<40(Type Ⅰ, Ⅱ) <60(Type ⅡR)	<49Pa or 5mm H_2O/cm^2(Level 1) <58.8Pa or 6mm H_2O/cm^2(Level 2, 3)	<49	<49
Synthetic blood penetration/kPa	120mmHg(ISO 22609) 16kPa(Type ⅡR)	80mmHg or 10.7kPa (Level 1) 120mmHg or 16kPa (Level 2) 160mmHg or 21.4kPa (Level 3)	120mmHg or 16kPa	N/A
Microbial cleanliness/(CFU/g)	≤30	N/A	≤100	≤100

N95 respirator

N95 respirators, like surgical masks, are examples of PPE that are used to protect the wearer from airborne particles and from liquid contaminating the face. However, both the Chinese and USA Centers for Disease Control and Prevention (CDC) do not recommend that the general public wear N95 respirators to protect themselves from respiratory diseases, including coronavirus (COVID-19). Those are critical supplies that must continue to be reserved for healthcare workers and other medical first responders, as recommended by current CDC guidance.

An N95 respirator is a respiratory protective device designed to achieve a very close facial fit and very efficient filtration of airborne particles. Note that the edges of the respirator are designed to form a seal around the nose and mouth. Surgical N95 respirators are commonly used in healthcare settings and are a subset of N95 filtering facepiece respirators (FFRs), often referred to as N95s.

The similarities among surgical masks and surgical N95s are:

(1) They are tested for fluid resistance, filtration efficiency (particulate filtration efficiency and bacterial filtration efficiency), flammability and biocompatibility.

(2) They should not be shared or reused.

Most N95 respirators are manufactured for use in construction and other industrial type jobs that expose workers to dust and small particles. They are regulated by the National Institute for Occupational Safety and Health (NIOSH), which is part of CDC. However, some N95 respirators are intended for use in a health care setting. Specifically, single-use, disposable respiratory protective devices used and worn by health care personnel during procedures to protect both the patient and health care personnel from the transfer of microorganisms, body fluids, and particulate material.

N95 respirators can be purchased with or without exhalation valves (Fig.40.2). Exhalation valves are a critical component of industrial respirators. They are designed to permit minimal inward leakage of air contaminants during inhalation and provide low resistance during exhalation.

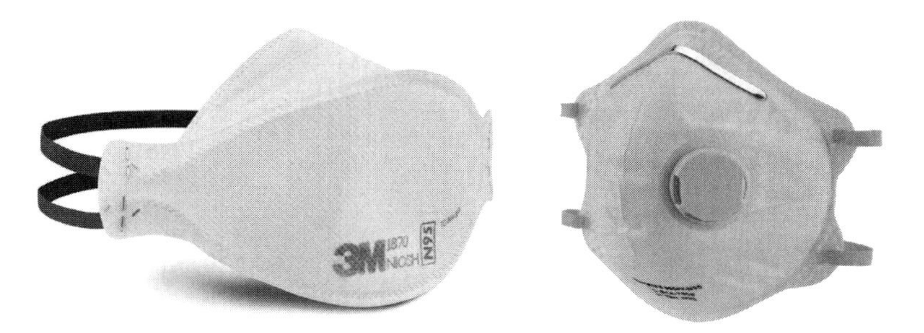

Fig.40.2 Photos of N95 respirator with (Right) and without (Left) an exhalation valve

Despite the comfort benefits of an exhalation valve for industrial workers, N95 respirators with exhalation valves should not be used when trying to maintain a sterile environment in an operating room. Although the exhalation valve makes breathing easier, any bacteria or virus expelled from the user may travel through the exhalation valve and enter the operating room, compromising the sterile environment.

Gowns

Gowns are identified as the second-most-used piece of PPE, following gloves, in the oridinary healthcare setting. Isolation gowns are defined as the protective apparels used to protect healthcare workers and patients from the transfer of microorganisms and body fluids in patient isolation situations. It is also defined as a gown intended to protect healthcare patients and personnel from the transfer of microorganisms, body fluids, and particulate material.

Historically, isolation gowns are used as a cover in isolation cases to protect the healthcare workers from the transfer of microorganisms and were made of 100% cotton or 50/50 cotton/polyester. Isolation gowns found in the marketplace today are produced from a variety of fabrics and a wide range of fibers. Isolation gowns are generally classified as "disposable/single-use" or "reusable/multi-use". In the U. S., disposable isolation gowns are used more commonly, while in Europe the share of reusables is larger. Approximately 80% of hospitals in the U. S. use single-use gowns and drapes.

Disposable (single-use) isolation gowns are designed to be discarded after a single use and are typically constructed of nonwoven materials alone or in combination with materials that offer increased protection from liquid penetration, such as plastic films. They can be produced using a variety of nonwoven fiber-bonding technologies (thermal, chemical, or mechanical) to provide integrity and strength rather than the interlocking geometries associated with woven and knitted materials. The basic raw materials typically used for disposable isolation gowns are various forms of synthetic fibers (e. g. polypropylene, polyester, polyethylene). Fabrics can be engineered to achieve desired properties by using particular fiber types, bonding processes, and fabric finishes (chemical or physical treatments). Reusable (multi-use) gowns are laundered after each use. Reusable isolation gowns are typically made of 100% cotton, 100% polyester, or polyester/cotton blends. These fabrics are tightly woven plain weave fabrics that are chemically finished and may be pressed through rollers to enhance the liquid barrier properties. Reusable garments generally can be used for 50 or more washing and drying cycles. Some of the characteristics of an ideal gown listed are: barrier effectiveness, functionality or mobility, comfort, cost, strength, fit, time to don and doff, biocompatibility, flammability, odor, and quality maintenance.

New Words and Expressions

surgeon 外科医生
pathogen 病原体，病菌
healthcare workers 医护人员
first responde 现场急救人员
face masks 口罩
gowns 手术衣，隔离衣
protective clothing 防护服
goggles 护目镜
respirators 呼吸器，呼吸机，呼吸面罩
airborne 空气传播的
aerosols 气溶胶
polystyrene 聚苯乙烯
polycarbonate 聚碳酸酯
polyethylene 聚乙烯

spunbond process 纺粘工艺
conveyor 传送带
electrets 驻极体
nose strips 鼻夹
ear loops 耳绳，耳带
influenza 流行性感冒
flammability 阻燃性
exhalation valves 呼吸阀

Notes to the Text

1. personal protective equipment (PPE)：指个人防护装备或用品。
2. Coronavirus Disease 2019 (COVID-19)：世界卫生组织(WHO)命名的专有名词，专指2019年新型冠状病毒。
3. resume work and production：复工复产。
4. China exported 224.2 billion masks between March and the end of December in 2020 to assist the international community in fighting COVID-19：中国在2020年3月到12月间共为国际社会抗击新冠肺炎疫情提供了二千二百四十亿只口罩。
5. Chinese General Administration of Customs (GAC)：中国海关总署。
6. ultrasonically welds：超声复合。
7. facial fit：面部吻合。
8. Centers for Disease Control and Prevention (CDC)：疾病预防控制中心。
9. National Institute for Occupational Safety and Health (NIOSH)：美国国家职业安全健康研究所。
10. Reusable garments generally can be used for 50 or more washing and drying cycles：可重复使用隔离服一般可以通过水洗或干洗的方式循环使用50次或更多。

Questions to the Text

1. Describe the component of a surgical mask.
2. What are the differences between the surgical masks and N95 respirators?

语言微课堂

Lesson Forty-One

Other Applications of Nonwoven Fabrics*

专业微课堂：
非织造材料的应用

Personal hygiene products

The good function and cost effectiveness of nonwovens that contribute to the protection, comfort and security given by sanitary protection products has insured their acceptance by the consumer. The products are disposable baby diapers, feminine hygiene products, and incontinence devices such as diapers and pads for adults.

Nonwovens used in these products provide softness, breathability, dryness and good wicking for a comfortable, absorbent product that protects against leakage. The absorbent core has a sophisticated designing to give an efficient structure that absorbs, disperses and retains fluid. In many product sample pad of wood pulp has been replaced by mixtures containing superabsorbent materials and synthetic fibers, for improved capacity to absorb fluids, and layers of nonwoven and embossed channels to direct fluids for more efficient absorption.

Breathable films and fabrics

These fabrics and films have permitted a new dimension in protection with comfort-in sports clothes such as skiwear and football uniforms, and in synthetic leathers for jackets, coats and shoes. Large and small pores can be formed in films made from poly(tetrafluoroethylene), and polyurethane to obtain breathability. A variety of processes have been used. If a film is embossed then stretched, holes will be generated. Large holes are also made with a high velocity water jets. Small pores can be created by dissolving out soluble particles or by evaporating volatile substances from fabrics and films. Small pores can also be punched out with laser beams.

A hydrophobic inner layer such as polyester or polypropylene can be combined with a hydrophilic outer layer such as cotton, rayon or hydrophilic nylon. Composites pull perspiration from the skin to the outside, leaving the wearer dry and warm.

Civil engineering materials

The use of nonwoven fabrics in civil engineering applications has grown rapidly since their introduction in the early 1970s. The largest end use of these fabrics is asphalt overlay, in resurfacing old roads, where it minimizes cracking and thereby extends the life of the new surface. This is followed, in decreasing consumption of yardage, by stabilization (separation), filtration, pond underliners, erosion control and reinforcement.

Nonwovens play a significant role in many fields. Note, pond underliner refers to the cushioning put on top of the impervious plastic sheet to protect it from abrasion and puncture. The use of liners is now mandated by law.

Nonwoven fabrics combine high strength, excellent durability, and good permeability to enable new construction techniques that enhance drainage, control siltation, and reinforce highways and railroads. In addition, they resist chemicals, bacteria and fungi.

Nonwovens provide high permeability and good filtration by maintaining excellent fiber coverage even when elongated under load. Their survivability is often superior to wovens-due in part to their ability to stretch sufficiently to allow irregular surfaces to become fully seated and supported by the soil.

As the reinforcement, a nonwoven fabric gives comfortable support and strength in all directions to the aggregate. As a separator, it prevents the loss of aggregate into subsoil. As a permeable membrane, it permits water to pass freely, while preventing the passage of soil particles. The relative importance of each of these functions depends upon the application, the type of subsoil, traffic and water conditions. In dry areas, the nonwoven is needed merely for added reinforcement. In swampy areas and in subtropical and tropical climates, the roles are more complex. For example, where there are alternating dry and rainy seasons, fine soil particles are prevented from passing through despite the fact that water is being pumped both ways.

In highway, railroad, parking lot, airport runway, and other stabilization applications, the nonwoven fabric is rolled onto the subsoil. Aggregate fill is then dumped onto the fabric and compacted. The area even improves with traffic as the fill becomes more compact. Less aggregate is needed, which means substantial savings in construction and materials.

In drainage applications a trench is lined with fabric and filled with aggregate. The nonwoven fabric is lapped over the top of the aggregate and covered with top soil. The nonwoven prevents soil particles entering and clogging the drain, thereby eliminating the need for expensive, well graded aggregate and perforated pipe. Thus, labor and material savings are realized.

Other applications include protection of underwater cables, embankment and dam stabilization, and construction of golf courses and tennis courts.

Fabric reinforced cement laminates is a relatively new class of construction material. The nonwoven fabrics, used within the cement before it is formed, are needle-punched polypropylene, and spunbond polypropylene or polyester-all resistant to the highly alkaline cement. They have pore spaces large enough to permit complete penetration by the cement.

Nonwoven reinforcements are cheaper than wovens. They are about the same cost per volume as chopped fiber. The continuity of the nonwoven fabric permits construction methods not possible with chopped fiber reinforcement, e. g. , continuous wrapping layup for the production of pipes and inflation techniques for constructing thin shell building enclosures. The continuity of fabric and sometimes the fiber in the fabric, as in spunbonded nonwovens, also provides strength not possible with discontinuous chopped fiber.

New Words and Expressions

hygiene 卫生学
disposable 用可弃的
feminine 女性的，妇女的
incontinence 失禁，无节制
leakage 泄露
caustics 腐蚀剂
corrosive 腐蚀的
radioactive 放射性的
skiwear 滑雪服装
uniform 制服
polyurethane 聚氨酯
volatile 挥发性的
polypropylene 聚丙烯
civil engineering 土木工程
asphalt 沥青
yardage 使用费用
pond underliner 池底铺层
siltation 淤泥
cushioning 垫子
impervious 不可渗透的
mandate 授权，命令
drainage 排水
aggregate 混凝土粒料
subsoil 下层土
swampy 沼泽的
airport runway 飞机跑道
embankment 筑堤

Notes to the Text

1. cost effectiveness：物有所值。
2. embossed channel：利用热压产生的沟状条纹。
3. These fabrics and films have permitted a new dimension in protection：句子中的“dimension”的意思是“领域，范围”。
4. control siltation：控制淤泥。
5. by maintaining excellent fiber coverage：保持较大的纤维含量。
6. Their survivability is often superior to wovens-due in part to their ability to stretch sufficiently to allow irregular surfaces to become fully seated and supported by the soil：“survivability”可以译为“使用寿命”；“to allow irregular surfaces to become fully seated and supported by the soil”意思是“利用非织造布作为土工材料，其使用寿命优于机织布，部分原因是非织造布能提供足够的伸长使得其不规则表面可以与土壤完全贴伏并被支撑”。
7. fine soil particles are prevented from passing through despite the fact that water is being pumped both ways：非织造布阻止细土颗粒通过，但水可以双向通过。
8. a trench is lined with fabric and filled with aggregate：沟壕用织物铺衬，然后覆盖粒料。
9. well graded aggregate and perforated pipe：严格选择的粒料和壁上有孔的管道。
10. used within the cement before it is formed：在水泥固化之前将织物置入。

Questions to the Text

1. Why nonwovens are suitable for the personal hygiene products?
2. Describe the nonwoven usage in the protective field.
3. Describe the process in applying the nonwoven in highway.
4. How to apply nonwovens in the drainage?

Lesson Forty-Two

语言微课堂

Appearance, Maintenance and Durability of Fabrics

The components of a textile fabric—fiber type, yarn structure or fiber arrangement, fabric structure, finish, and color must all be considered in the selection, use, and care of a textile product. The relative importance of each of the factors is closely related to the appearance, maintenance, durability and comfort of the product. The relative importance of each depends upon the planned end use of the textile. For best results, it is important that the consumers' choice be similar to that identified by the manufacturer. Selecting a product for a drapery that the manufacturer intended for use in evening apparel could result in a very dissatisfied customer. To avoid problems of this type, the consumer should become familiar with the different factors involved in choosing textile products and make a selection based on known facts and on those that can be determined through the individual's knowledge of how to evaluate textiles.

Appearance

专业微课堂：
服装的清洁保养

The appearance of a fabric can be described as the visual effect on consumers, who, in addition, make use of the sense of touch to further comprehend what they see. Thus, both the visual and tactile senses are involved in appearance.

Fiber luster and texture influence the appearance of fabric. Shiny, lustrous fibers make it possible to make shiny, lustrous fabrics. Dull fibers may be made into fabrics with a dull appearance; however, it is possible to add finishes to dull fibers to enhance their luster. The yarn structure or fiber arrangement of a fabric influences appearance. Complex yarns produce a fabric with appearance characteristics that differ markedly from the surface obtained with uniform simple yarns. Yarns or fiber arrangement can be responsible for varying character through the special shapes and forms that characterize such yarns and fiber groupings. Fabric structure is important in appearance. Woven fabrics differ from knitted fabrics; both differ from nonwovens; lace and braid are still different. One can easily recognize different methods of constructing fabric, and each method creates special appearance qualities. Color is an obvious part of appearance and one of the most important factors in the selection, use, and care of the product. Finish may alter or enhance appearance. Specific attention given in the various sections of this text will enable the reader to discern how the components fit together in developing the appearance of fabric. Depending on the interaction of the various factors, fabrics may appear to be soft or stiff, rough or smooth, delicate or coarse, lustrous or dull, bulky or sheer, bright or dull, light or dark.

Maintenance

The maintenance given a textile product is dependent on the fiber or fibers used, yarn structure or fiber arrangement, fabric construction, method of imparting color, type of dyestuff used, finishes and methods of applications, and geometric factors.

Inherent and geometric fiber properties, yarn structure or fiber arrangement may affect maintenance. Such characteristics as the number of turns per meter and the type of construction of yarn—simple, complex, single, ply, cord all play a part. For example, a yarn with a very high number of turns per meter, such as crepe yarns, may shrink when subjected to moisture and undergo excessive shrinkage. A complex yarn can be damaged by abrasion from equipment during the maintenance process. Blended yarns should receive the type of care recommended for the more easily damaged fiber. However, proper blending may result in easier care for both fibers. For example, a blend of a thermoplastic fiber with cotton produces a fabric that is easily laundered, dries quickly, and requires lower ironing temperature than pure cotton. A blend of wool and acrylic can be home laundered, whereas dry cleaning is recommended for pure wool.

Proper laundering techniques should be observed for all washable fabrics. Instructions on care labels contribute specific knowledge and should be followed if durability is to be achieved. Consumers should recognize that many factors influence laundry results, including type of detergent used; laundry additives used; length of the wash cycle; type of bleaches selected in relation to the fiber content, dyestuffs, and finishes involved; water softness selected; and fabric softeners.

Color is an important aspect of care. Whether fabrics are dyed or printed, color may be lost during laundering or dry cleaning. When convenient, it is a good idea to determine colorfastness of dyestuffs before using a specific care procedure. As this is extremely difficult, if not impossible in many cases, most consumers are left with the care instructions on the label as the only guide. Finishes applied to fabrics may make maintenance or care easier, but they may also cause new problems. Some finishes respond well to dry cleaning but are destroyed by laundering; others respond to laundering but are destroyed by dry cleanings. Again, information on the care label should be studied carefully. Further, information concerning special finishes may be a source of valuable information concerning fabric care.

Durability

Durability is defined as the ability to retain properties and characteristics for a reasonable period of time. Factors involved in durability are also important to appearance, as retained appearance depends on the durability of the textile. The following generalizations concerning durability are based on the component parts of any textile fabric.

Fiber content influences durability. Some fibers, such as acetate, are valued for their beauty of hand and drape rather than for their wearing qualities, whereas, nylon is frequently selected because of its strength, abrasion resistance, and other properties that contribute to durability. It is important to study each fiber group and identify the properties that contribute to durability, such as relative tenacity and flexibility. Strong fibers are more durable than weak fibers if other factors are held constant. Fibers with good flexibility are more durable in apparel than stiff fibers because of the bending that is required, but a stiffer fiber may be better for some home furnishings where it could resist crushing and retain shape. It is, once again, important to identify the end use of the product and select the fiber or fibers that will give good durable service in that specific end use.

Yarn structure or fiber arrangement determines the durability of a textile to a degree. In general, fabrics made with yarns tend to have greater durability than nonwovens that have been made with fiber arrangements. This is particularly true for fabrics designed for apparel.

Fabric structure is an important factor in durability. Plain and twill weave fabrics are more durable than satin weaves; the floating yarns in a satin weave are subject to snagging, breaking and damage from abrasion. Filling or weft knit fabrics of plain designs are subject to runs when the yarn is broken.

Felt fabrics usually have good durability unless they are subjected to considerable pulling force or abrasion. Nonwovens may be durable or disposable, depending on the planned end use. Nonwoven interfacings should be able to withstand various methods of maintenance to give satisfactory life in the final product.

In general, basic weave constructions and plain knit fabrics are more durable than complicated fabric constructions, which could show signs of wear as a result of surface distortion or damage.

Color selection and method of application influence durability. Choosing the proper dyestuff for each fiber type is essential if the fabric color is to give good service. The consumer, however, should be aware of scientific, economic, and fashion affects on color. It is impossible to find all colors in all classes of dyestuffs, and this may mean selection of a second choice dye in order to obtain a currently popular or fashionable color. Moreover, not all dyes within any one class are equally good, and it is difficult to find dyes in all colors that prove to be color fast to all degrading environmental conditions.

Finishes, whether applied for appearance or to alter certain behavioral properties, often affect fabric durability. Surface finishes such as glazing can be removed by improper care, so that the useful life of the product is reduced. Most wrinkle recovery, minimum care, or permanent press finishes reduce fabric tear strength, breaking load, elongation, and abrasion resistance. However, minimum care finishes, despite reduced durability, are frequently more attractive than comparable fabrics that do not have such finishes. This may compensate many consumers who are willing to accept a shorter use life for such products.

Durability is determined by several factors. Consumers should determine the relative importance of durability in terms of product end use and make their selection accordingly.

New Words and Expressions

visual effect 视觉效果
tactile sense 触感，触觉
crepe yarns 绉纱
shrink 收缩
run 针织物脱散
interfacing 界面
glazing 轧光
care 保养，护理

Notes to the Text

1. Selecting a product for a drapery that the manufacturer intended for use in evening apparel could result in a very dissatisfied customer：工厂生产的用于晚装的面料，如果顾客用做窗帘，肯定会不满意。
2. all play a part：都会起作用。
3. A blend of wool and acrylic can be home laundered, whereas dry cleaning is recommended for pure wool：意思是毛/腈混纺织物的洗涤方式比纯毛织物要简单，可以手洗，而纯毛织物则建议干洗。
4. through the special shapes and forms that characterize such yarns and fiber groupings：通过这些纱线和纤维特殊的形状和结构。"that characterize such yarns and fiber groupings"意思是"它们是这些纱线和纤维的特色"。
5. As this is extremely difficult, if not impossible in many cases, most consumers are left with the care instructions on the label as the only guide：在许多情况下这样做是十分困难的，多数消费者只能依靠说明书。
6. are valued for their beauty of hand and drape rather than for their wearing qualities：更多地由于其良好的手感和悬垂性而受到人们喜爱，而不是其穿着品质。
7. and this may mean selection of a second choice dye in order to obtain a currently popular or fashionable color："second choice dye"意思是"(色牢度)属于二档的染料"。
8. degrading environmental conditions：引起褪色的环境条件。

Questions to the Text

1. What factors may influence the fabric appearance?
2. In taking care of a textile product, what should be taken into consideration?
3. How to take care of blended textiles?
4. What factors may influence the laundering results?
5. How does the fiber influence the durability?
6. What is the relationship between yarn structure and fabric durability?
7. Why fabric structure is an important factor as far as fabric durability is concerned?
8. How does finishing affect the fabric durability?
9. Why color selection is involved in discussing the fabric durability?

Lesson Forty-Three

Fashion Style and Design

语言微课堂

专业微课堂：
服装时尚

The word style was used in our definition of fashion. Sometimes, one hears the words style and fashion used interchangeably, but they do have different meanings and should not be considered synonymous. Style is a particular characteristic or feature that distinguishes one object from another. In apparel, a style is a particular characteristic of design or silhouette. Skirts and pants are examples of different styles, and each comes in different designs that may also be considered styles. For example, skirts are designed in many styles, such as A-line, tubular, dirndl, flared and pleated. Pant styles include trousers, Bermuda shorts, pedal pushers and bell bottoms. Styles remain the same, whereas fashions change. A style is always a style, but it is in fashion only as long as it is purchased and worn by a significant number of consumers. The distinctive cut or style of a blazer jacket remains the same, but it will not always be in fashion. Fashion designers reintroduced the miniskirt in 1987, but the style was not widely accepted by a majority of consumers. Because of this rejection, sales in many stores were lower than predicted, and large price reductions were taken by retailers in an attempt to sell the goods.

Home furnishings also are available in different styles. Basic areas of furniture styles are traditional, country or provincial and contemporary. There are variations within each general style. Style is also used to describe people. Certain individuals are said to have style, meaning that they have a distinctive manner setting them apart from others. Some well-known individuals considered to have distinctive styles.

The word style has yet another meaning within the fashion industry. A manufacturer uses a style number to identify each item in his or her line. When retail buyers order or reorder an item in a manufacturer's line, the style number is listed on the order form. Thus in this usage the word style refers not to the particular style of the item but rather to the number given each item in the line.

A design is a unique version of a style. For example, a shirt waist is a specific dress style, but it can be interpreted in numerous ways. Each interpretation or version is a design. Fashion design involves the manipulation of four basic elements: form or silhouette, detail, color and texture. By changing these elements, one develops a new style that may become a fashion.

Silhouette is the overall outline or shape of a costume. It is also referred to as contour or form. Three basic silhouettes have been worn by women throughout the history of fashion. These three basic silhouettes are ① the straight or tubular, ② the bell shaped or bouffant, and ③ the bustle.

Many variations of each of these silhouettes are possible. Details are the various components within a silhouette such as collars, sleeves, shoulder and waist treatments, and length and width of skirts and pants. Trims used to decorate the garment are also part of the detail. Trims include buttons, lace edging, appliques, embroidery. Great variety in designs can be achieved through varying the detail.

Texture refers to the look or feel of a fabric or other construction material. It creates surface interest and effects the sensuous element of design. Texture greatly influences the appearance of the silhouette and changes the mood of the garment. The drape of the garment can change significantly with a change in fabric texture. A dress made in a soft flowing fabric such as silk chiffon will look very different from the same dress style constructed in a stiff taffeta fabric.

Color is often the most important element in a design. Highly influenced by both fashion and customer preference, color is usually the first factor the customer considers when selecting a garment. A change in color obviously makes a considerable change in the appearance of a garment. Color can make the garment and its wearer look larger or smaller. Color can also affect the mood of a garment, making it appear informal or dressy.

New Words and Expressions

fashion style　时装款式，样式，风格
A-line　A 字形裙
silhouette　侧面影像，服装轮廓
tubular skirt　筒裙
dirndl skirt　紧身连衫裙
flared skirt　喇叭裙
pleated skirt　百褶裙
pants　裤子
trousers　裤子
Bermuda shorts　百慕大短裤
pedal pushers　（长及小腿的）女运动裤，自行车裤
bell bottom　裤子喇叭脚口
blazer jacket　运动夹克，宽松外衣，便装
miniskirt　超短裙
shirt waist　女用衬衫
bouffant　蓬松裙
bustle　裙撑
applique　嵌花
embroidery　刺绣
mood　服装的情调
silk chiffon　薄绸
dressy　衣着讲究的

Notes to the Text

1. The word style was used in our definition of fashion：句中的“was”在翻译时应体现出来，即译为“曾经……”。
2. A style is always a style, but it is in fashion only...by a significant number of consumers：句子中“but it is in fashion”意思是“但要称得上时装”。
3. sales in many stores were lower than predicted：“sales”译为“销售额”。
4. Basic areas of furniture styles are traditional, country or provincial, and contemporary：“country or provincial”直译为“县或省”，在这里可译为“地方型”。

5. Certain individuals are said to have style, meaning that... apart from others：句中从“meaning”直到句子结束这部分做整个句子的状语。
6. Thus in this usage the word style refers not to the particular style of the item but rather to the number given each item in the line：句中“refers not to...but rather to...”意思是“不是指……而是指……”。
7. Highly influenced by both fashion and customer preference：做主句的条件状语，意思是“色彩受到时装和消费者喜好的影响”。
8. making it appear informal or dressy：“informal”的意思是“非正式的”，在这里指穿着随便的便装风格，译为“潇洒的”比较合适。

Questions to the Text

1. Say something about the fashion style.
2. Give examples of various styles for various things.
3. What are the details of a dress?
4. Why texture is important for a garment?
5. Give reasons to the importance of color in a design.

Lesson Forty-Four

Fashion Movement*

语言微课堂

专业微课堂：西方百年服装发展史

Several fundamental principles of fashion serve as a foundation for fashion identification and movement. These fashion principles remain constant; fashions change, but these principles do not. They are a solid foundation for identifying and predicting fashion trends. The basic principles of fashion movement include the following.

(1) Fashion movements are usually evolutionary in nature and are rarely revolutionary. Fashions usually evolve gradually from one style to another—evolutionary fashion rather than changing rapidly. This is often illustrated in the changing length of women's skirts. Seldom is a major change in skirt length accepted in one season. Instead, skirt length usually inches up or down gradually over a period of several seasons or even years. In the late 1950s and during the subsequent decade, skirt length began moving upward approximately an inch a year until skirts reached the mini style of the late 1960s. Throughout the 1970s skirt length gradually became longer.

(2) The consumer makes fashion. It is the consumer, not the designer or manufacturer, who determines what will be in fashion by accepting one style and rejecting another. Although designers, producers and retailers can encourage or retard the process of new fashions, it is consumers who ultimately determine fashion acceptance.

(3) Price does not determine fashion acceptance. The acceptance of fashion does not depend on price. Although a new style may originate at a high price, it can quickly be made available at various prices. A custom-made designer dress may be priced at several thousand dollars; however, once the style is copied in ready-to-wear, a variety of prices are available to consumers. Sometimes copies are mass produced so quickly that they reach the stores even before the original designs. The trimmings and fabric may be different and the workmanship may not be equal in quality, but the dress style is basically the same.

(4) Fashion movement is not dependent on sales promotion. Promotional activities, such as advertising, fashion shows and window displays, cannot sell merchandise that consumers do not want to buy. Although advertising and other promotional efforts by manufacturers and retailers help to generate sales, they do not dictate what styles will be accepted by consumers. Once a fashion trend is established, however, promotional efforts do aid development of the trend.

(5) Fashions often end in excess. Paul Poiret, a well-known French high-fashion designer of the 1910s, noted that "all fashions end in excess". This saying is true for fashions today as well.

Once a fashion has reached an extreme in styling, a new and different look will occur. For example, in the late 1960s women became tired of bouffant hairstyles and sought new looks. Another example is the fashion silhouette of the mid-1980s, which emphasized shoulder width. When shoulder pads began to make women look like football players, the fashion for extreme padding ceased.

Fashion is a major force in our daily lives. It affects every aspect of our lives, influencing what we see, do, and wear. We are influenced by fashion from the time we get up in the morning until we go to bed at night. Fashion affects the clothing and accessories we wear, the environment in our homes and offices, the car we drive, the food we eat, and the entertainment we enjoy. Fashion is everywhere, and it is unavoidable. Fashion is so integrated into our lives that we often are unaware of its impact. Some people say that they are not concerned with fashion. However, even these people are influenced by fashion even if only by trying to avoid it.

Men, women and children all feel the effect of fashion in the clothes they wear as well as in other aspects of their appearance. Fashion extends even to the family pet, which wear a fashionable collar or sweater and eat from an appropriately decorated bowl. Fashion encourages women to wear their hair short or long, curly or straight. Fashion influences the colors, the fabrics and the styles of women's clothing. It also determines the types and colors of cosmetics worn. Men, as well as women, feel the impact of fashion. Fashion leads men to grow beards or be clean shaven, to wear wide or narrow ties, to choose single or double breasted suits. In recent years fashion has encouraged men to add more color to their wardrobes. At one time white was the only acceptable color for a man's dress shirt. Today the dress shirt is available in a variety of solid colors and stripes. Men's sportswear, too, has become almost as colorful and fashion oriented as women's sportswear.

Children also respond to fashion. Clothing designed for preschool age children often follows the fashion for adults. Children are influenced by their peers beginning at an early age to dress as others dress. They are often teased and ridiculed if their way of dressing is different from that of their peers. The influence of fashion grows stronger as the teen years approach. Wearing the right brand name clothing is often very important to the teenager.

New Words and Expressions

mini style 服装的超短款式
custommade 定制的，定做的
trimming 服装的装饰与修剪
midi 半长裙
shoulder pad 垫肩
accessories 服装的附件
collar 领圈
cosmetic 化妆品
single breasted 单排纽扣的
double breasted 双排纽扣的
peer 同等的人

Notes to the Text

1. fashions change, but these principles do not：在“do not”后面省略了“change”，该句的意思是“时装在变化，但这些原理不变”。
2. Instead, skirt length usually inches up or down gradually over a period of several seasons or even years：“inches up or down gradually”意思是“逐渐而缓慢地变化”。
3. skirt length began moving upward：如果直接译为“裙子的长度开始向上移动”，会令人难以理解；实际的意思是“裙子的长度开始变短”。
4. Price does not determine fashion acceptance：句中“acceptance”如果翻译为“接受”，则句子显得不通顺，根据上下文，可译为“价格并不影响时装的推广”。
5. ready-to-wear：意思是“成衣”，在此处译为“批量生产”。
6. When shoulder pads began to make women look like football players：“football”在美国多指橄榄球，在美国足球一般被称为“soccer”。
7. It affects every aspect of our lives, influencing what we see, do, and wear：句中“influencing what we see, do, and wear”为状语从句，说明时装是如何影响我们生活的。
8. It also determines the types and colors of cosmetics worn：句中“worn”的原型是“wear”，意思是“穿戴，使用”，它做“cosmetics”的后置定语。
9. In recent years fashion has encouraged men to add more color to their wardrobes：句子中“to add more color to their wardrobes”可译为“选用多花色的服装”。
10. teen years：严格讲指13~19岁的一段时间，因为英语单词从第十三到第十九，均以“teen”结尾，这里泛指十几岁的孩子。

Questions to the Text

1. Why the fashion movements are usually evolutionary in nature?
2. How does the consumer make fashion?
3. How do you think of “everything ends in excess”?
4. Why fashion influences our daily life?
5. How do you feel the influence of fashion trend?

语言微课堂

Lesson Forty-Five

Mechanical Properties of Fabrics

专业微课堂：
织物拉伸强力
测试原理

The properties of any textile product are governed by a variety of factors. One determining factor is the geometrical relationships that exist among the component parts of a textile; fiber, yarn structure or fiber arrangement, fabric structure, finish, and color. How these components are assembled influences the properties of the textile fabric, especially the geometrical arrangement of fibers into yarns and fabrics.

Consider the component parts and how each one can vary; then consider how such variations influence the final fabric. Fiber properties are determined by the inherent characteristics of the base polymer and such geometric features as fiber length, cross section shape and area, amount of crimp or texturization, stiffness or softness, and surface contour. Yarn properties are influenced by fiber characteristics and such geometric features as amount of yarn twist, shape and diameter, compactness of fibers within the yarn, and the basic yarn structure, simple, complex, single, ply, cord.

Fabric properties are the result of fiber and yarn characteristics or the method of arranging the fibers in making the fabric plus the special geometry of the fabric structure itself. This includes the type of fabric structure, the size of yarns used, the number of yarns per 10cm in fabrics, the number of loops per 10cm in knits, the fiber arrangement in nonwovens, and the presence of dyes and finishes that influence the behavior of the fiber, yarn, and fabric.

Any fabric should possess adequate resistance to tearing and stress. Fabrics that are made from inherently strong fibers will tend to have adequate strength for the selected use and care. However, other factors may influence the properties, and the fabric may be subject to quick wear because of low resistance to tear and pulling stress.

Combined geometric features that produce a highly durable fabric that resists both tearing and pulling stress are found in fibers such as nylon, which has inherently high tensile strength; filament fibers, which tend to possess better resistance to tear and tensile stress; yarns of ply construction; and a plain weave fabric structure with sufficient yarns to provide good strength but not so many as to make the fabric stiff. Changing one factor can reduce the combined geometric properties and reduce the resistance to tearing or pulling forces.

Tearing strength can be increased by using strong fibers in strong yarn constructions in both warp and filling directions of a woven fabric and by general use of knitted structures. However, tearing strength is also affected by yarn mobility within the structure. High tearing strength can be achieved when yarns that have relatively low strength are combined into a fabric using a loose open structure. Low tearing strength can result when strong yarns are combined into a compact structure. As yarn mobility is decreased, tearing strength may be decreased.

Tensile strength can be improved by using strong fibers and yarns in both directions in a woven fabric and in all yarns for a knitted fabric; however, if the yarns in one direction of a woven fabric, or alternate yarns of reduced strength in a knit, are used, the resulting fabric may be relatively weak in the one direction and be subject to damage from a small amount of pull or stress in that particular situation or direction. A fabric that has been constructed with 400 yarns per 10cm in both the warp and filling will be stronger than a fabric made of 400 yarns in the warp and only 300 in the filling—provided that the yarn size is consistent between both fabrics. Conversely, a fabric made of 400 yarns per 10cm in both warp and filling may be considerably weaker than another fabric made of the same number of yarns but composed of fibers that are inherently stronger.

Further, a fabric with 400 yarns per 10cm in each direction but of yarns that are smaller in diameter and with low levels of twist would be considerably weaker than a fabric with 400 yarns per 10cm but of a heavier diameter or denier and with medium to high twist.

Tensile and tearing strength can also be influenced by the type of yarn used. A single or simple ply yarn in which the fibers and twist are uniformly distributed will probably give better resistance to stress than novelty or complex yarns where there may be areas of reduced strength owing to uneven twist, uneven distribution of fiber, and uneven yarn diameter.

Balanced woven fabric constructions will tend to resist tearing or pulling stress more successfully than fabric of unbalanced construction, as the yarn distribution is equalized in the balanced fabric and thus strength is equalized. Knit fabrics that are made of yarns of the same size but with a different number of loops per 10cm will differ in their resistance to pulling force. Obviously, the more yarns, the higher strength—up to an optimum point.

Nonwoven fabrics will resist tear or pulling stress best when the fiber arrangement is spread over all directions and when the method of sealing the fibers together resists pull or tearing stress.

The presence of finishes will influence tear and tensile strength of fabric despite the strength created by the choice of fibers, yarns and construction processes. Regardless of the geometric factors at work, certain finishes do reduce resistance to tearing and pulling. The use of easy care or durable press finishes reduces the resistance of a fabric to tearing stress; consequently, in making good durable press fabrics it is important to select the most optimum geometric factors in order to provide a fabric with adequate resistance to damage from tearing—the most seriously affected property by the presence of such finishes.

If consumers keep in mind that fibers provide the most inherent strength, types of yarn or fiber arrangement that contribute good resistance to stress, and a fabric construction that is well balanced and uniform, they may be able to choose a fabric that exhibits outstanding resistance to damage from either tearing or pulling force.

New Words and Expressions

texturization　纱线变形工艺
crimp　卷曲
stiffness　刚度、硬度
compactness　紧密度
tearing　撕裂
pulling stress　拉伸应力
tensile stress　拉伸应力
tearing strength　撕裂强度
tensile strength　拉伸强度
uneven　不均匀的
balanced fabric　平衡织物，经纬向参数相同的织物
sealing　黏结

Notes to the Text

1. and such geometric features as amount of yarn twist, shape and diameter：还有诸如捻度、纱线形态、纱线直径等这样的几何特性。
2. single, ply, cord：这里指单纱、合股线和绳线。
3. yarn mobility：织物中纱线的可移动性。
4. the more yarns, the higher strength—up to an optimum point：在一定范围内，纱线数量越多，织物强度也越大。"optimum point"为最佳点，超过该点，这种关系可能就不存在了。

Questions to the Text

1. What are the major factors that influence the tearing and tensile strength of the fabric?
2. Give the necessary conditions of a fabric with best tensile strength.
3. How is the tensile and tearing strength influenced by the type of yarn?
4. Give an example of balanced woven fabric.
5. Name all the yarn types you know.

Lesson Forty-Six

Other Physics Properties of Fabrics*

Thermal properties

Fabric for use in selected items of wearing apparel and home furnishings must have properties that make it adaptable for various climatic conditions. The end use of a fabric may require that the fabric hold heat against the body, prevent heat from penetrating to the body, or conduct heat away from or toward the body.

Heat can be lost from the body by conduction, convection, radiation, and evaporation of perspiration. Fabrics used in the cold environments should entrap and hold body heat. They should prevent heat loss by conduction, convection, or radiation. To reduce heat loss by conduction, fibers should be combined into yarns and fabrics in such a manner that air spaces are left to serve as insulating areas.

To prevent heat loss by convection and to some degree, by radiation, the following features would be desirable: ① a fabric with smooth yarns that are packed densely, and ② a high thread or fabric count with very tiny interstices between yarns to reduce air permeability.

There are measurements that help to identify thermal properties of fabric.

Clo identifies a unit of thermal resistance defined as the insulation required to keep a resting person comfortable in an environment at 21℃, air movement of 0.1 meter per second, or roughly the insulation value of typical indoor clothing. The thermal resistance of a fabric is identified as the *R* value. These values may be determined scientifically and provide some indication of the thermal properties of fabrics or assemblies of fabric. It has been suggested that the total apparel for wear on a hot day should not exceed 0.5 clo, but on a cold winter day in cold climates a person may need 4.0 to 5.0 clo to be warm. Clo values depend on the ability of a fabric or layers of fabric to control the movement of air through them.

In warm areas clothing should accelerate or at least aid in heat loss. Where a minimum of insulation is desired, fabrics of loose weaves, smooth yarns, fine diameter yarns, and large interstices will be wise selections. These fabrics contribute to heat loss by convection, for air currents can circulate freely; by radiation, since the open spaces permit the rays to move to the surrounding air; and by evaporation, because the moisture vapor can move easily into the air. If the fibers are good conductors and if the fabrics have few or no spaces to entrap air, heat loss by conduction is aided.

In regions where air temperatures exceed 37℃, particularly desert and tropical regions, fabrics are sometimes chosen that prevent passage of heat toward the body. These fabrics are similar to those recommended for cold climates, and the purpose is to hold the high temperature out instead of letting it pass through the fabric to the body.

The air permeability is closely related to the thermal properties and is frequently a major factor in body comfort and in protection against moisture buildup as well as heat retention or release. Fabrics with good air permeability encourage heat loss by air movement. As air freely moves through a fabric, heat can be dissipated to the outside; conversely, heat can pass through the fabric to the body.

The air permeability of a fabric is influenced by several factors: the type of fabric structure, the design of a woven or knit structure, lacy structures versus compact weaves or knit, the number of warp and filling yarns per 10cm, the number of loops per 10cm in knit fabrics, the amount of twist in yarns, the size of the yarns, the type of yarn structure, and the size of the interstices in the fabric. Fabric with low thread counts and fine yarns usually have good air permeability. Whereas compact fabrics with high thread counts and very tiny interstices between yarn interlacings or interloopings will naturally limit the passage of air. In addition to influencing heat loss by convection, air permeability may be related to the movement of moisture or water vapor adsorption, and/or wicking properties. To make comfortable hot weather apparel from hydrophobic fibers, fabrics should be of a porous construction so that they permit air permeability, moisture diffusion, and wicking.

Moisture properties

The movement of moisture into fibers or along the surface of fibers, yarns and fabrics is the result of wicking, moisture absorption, and moisture regain properties. Hydrophilic fibers will absorb moisture and allow it to pass through geometric openings in the fabric if those openings are of sufficient size. Hydrophobic fibers do not absorb moisture; however, these fabrics provide for moisture transmission by geometric openings, by adsorption onto the surface or between fibers, and by wicking. Both hydrophilic and hydrophobic fabrics allow moisture transmission if the fibers have a tendency to wick.

The movement of liquid moisture and vapor through fabric is dependent on the compactness or looseness of the weave or fabric structure, the yarn structure or fiber arrangement, and the wicking and adsorption characteristics. When fabrics are composed of hydrophilic fibers, vapor will move through both the fibers and the fabric interstices. Hydrophobic fibers permit the passage of moisture vapor through the interstices, between fibers within yarns, and along fibers and yarns.

The transmission of liquid moisture may be enhanced by loose fabric structure; however, liquid moisture moves comparatively well through a porous structure even though that structure may be compact, provided the fibers do have wicking properties. Wearer comfort and discomfort appear to be closely allied to the ability of a fabric to pick up water by means of wicking, adsorption(adherence of moisture to the surface of the fiber), or absorption (moisture actually entering the fiber). In addition, the fabric should provide for some water vapor transmission. Fabrics constructed of hydrophobic fibers may be uncomfortable if the geometric features of the fabric are not properly planned and if wicking is not present or is hindered in some way.

Blends that include cellulosic fibers like cotton are especially good in water transmission because cotton fibers have a high degree of wickability as well as absorption. Other cellulosic fibers also improve moisture passage. They are extremely efficient in picking up water from the skin surface and carrying it into the fibers within the yarns and fabrics.

The type of fiber, filament or staple, is an important aspect of comfort. Studies indicate that filament yarn fabrics tend to plaster themselves to the skin when liquid moisture is present, whereas spun or staple fiber yarns are held away by the minute fiber ends extending from the surface.

Water repellency is determined to some degree by the geometry involved in relation to moisture characteristics. Fabrics with large interstices will permit rapid transfer of water, but fabrics constructed with very tiny interstices tend to repel water. This behavior can be enhanced by using fibers that are hydrophobic, nonwicking, and nonabsorbent and have a high surface tension.

Fabrics made of soft cotton yarns can provide good water repellency by virtue of their ability to absorb considerable moisture and swell. As the fibers swell, so do the yarns, the interstices are closed, and transmission of moisture is reduced.

Sometimes a high degree of water absorption is desired in a fabric. The geometric features for such fabrics are altered to provide a maximum surface of soft yarns with good absorptive properties. For example, terry cloth is frequently selected for its absorbency. The geometry of this fabric includes relatively dense construction with yarns that are soft and loosely twisted. Furthermore, the loops of yarn tend to hold moisture within the pile until it moves into the yarns and base fabric structure.

Wrinkle recovery

专业微课堂：织物褶皱回复性测试

Fabric geometry is important in relation to wrinkle recovery. The following points indicate some of the interrelationships between wrinkle recovery and geometric features. Very short fibers tend to be displaced easily when yarns are folded and will therefore retain permanent deformation including wrinkles and creases. Medium length staple fibers in relatively firm yarns, long staple fibers in medium firm yarns, and filament fiber yarns produce good wrinkle recovery if properly combined into yarns and fabrics.

Actual fiber diameter and shape influence wrinkle recovery. Fibers that are round in cross section tend to resist bending and folding. Fibers that resist bending and folding usually recover rather quickly from light to medium folding stress. However, if sharp, heavy wrinkles are formed in round fibers, they recover slowly because the factor of strain tends to hold the creases in place.

The amount of yarn twist contributes to wrinkle recovery. Yarns of medium twist provide little or no opportunity for fiber displacement, so these yarns tend to return to their original position. Yarns with very low twist permit fiber displacement when bent or folded, and, upon release from strain or stress, the fibers do not return to their original position. High twist yarns do not recover as well as medium twist yarns because the high twist tends to hold folds or ceases as a result of strain or stress on the fiber and/or the yarn.

Coarse yarns composed of fine staple fibers or filaments resist wrinkling and, if wrinkled, recover more effectively than yarns made of coarse fibers, filaments, or monofilaments.

Woven fabrics of basket, twill, or satin weave constructions recover more easily from wrinkling than plain weave fabrics if the fabrics are of equal compactness and yarn size. The basket, twill and satin weaves have higher yarn mobility than the plain weave because they have fewer yarn interlacings. In general, the lower the number of yarns per 10cm the better the wrinkle recovery.

A compact twill weave fabric with a relatively high thread count made of fine yarns will probably have a lower wrinkle recovery than a coarse plain weave made from larger yarns. However, in general, any type of fabric, regardless of the weight or tightness of construction, would give better easy care performance if constructed from weaves with fewer interlacings than the plain weave.

New Words and Expressions

conduction 传导
convection 对流
physiological 生理的
vaporization 汽化
thermal resistance 热阻
clo 克罗值(服装保温能力单位)
thermal resistance 热阻
wicking 芯吸
moisture absorption 吸湿性
moisture regain 回潮
hydrophilic fibers 亲水纤维
hydrophobic 疏水的，拒水的
water repellency 拒水性，疏水性
wrinkle recovery 褶皱回复，折痕回复
fiber displacement 纤维位移
diffusion 扩散
adsorption 吸附，吸收
monofilament 单丝

Notes to the Text

1. may require that the fabric hold heat against the body: "hold heat against the body"意思是"使热量不接近身体"。
2. Clo identifies a unit of thermal resistance defined as the insulation required to keep a resting person comfortable in an environment at 21℃, air movement of 0.1 meter per second: 一个安静坐着的人，在室温21℃(相对湿度小于50%)，风速不超过0.1m/s的环境中感觉舒适时，所穿衣服的热阻值为1克罗值。
3. lacy structure: 网状结构（指松散的结构）。
4. provided the fibers do have wicking properties: "provided"意思是"倘若，假如"。

5. filament yarn fabrics tend to plaster themselves to the skin when liquid moisture is present: "to plaster themselves to the skin"意思是"紧贴在皮肤表面"。
6. High twist yarns do not recover as well as medium twist yarns because the high twist tends to hold folds or ceases as a result of strain or stress on the fiber and/or the yarn: "as well as"在本句子中的意思是"像……那样"。

Questions to the Text

1. How many ways are there for the heat to loss from body? And give short explanation to each.
2. What is the relationship between air permeability and thermal properties of the fabric?
3. What kind of clothing is considered good in cold climate?
4. To design a fabric used in summer, what are the main factor to be considered?
5. How does the fabric transmit liquid moisture?
6. What is the difference between absorption and adsorption?
7. What fabric would have an excellent wrinkle recovery?
8. How is the terry cloth constructed?
9. How does the fiber diameter and shape influence the wrinkle recovery?

语言微课堂

Lesson Forty-Seven

Standards for Textiles and Apparel

It could be about making a product, managing a process, delivering a service or supplying materials—standards cover a huge range of activities. Standards are the distilled wisdom of people with expertise in their subject matter and who know the needs of the organizations they represent—people such as manufacturers, sellers, buyers, customers, trade associations, users or regulators.

Standards and Testing are crucial to providing quality service, ensuring safety and reliability, and creating customer confidence. The testing and certification industry has been playing an increasingly significant role due to more stringent environmental and safety regulations.

The textile and apparel industry supports a variety of related sectors, from fibers to clothes, from home furnishing fabrics to industrial textiles. Every kind of textile-based product can be produced in a countless number of ways and comes in a variety of styles and functions.

Textile producers have a strong incentive to take part in developing their own standards for measuring their products, with regard to the product's specific purpose and its ability to stand up to repeated use. Production companies are motivated by their long-term relationship with their customers, and often they will use their commitment to verifiable quality standards as part of their marketing strategy.

For a number of reasons, every proposed product and its use must be appropriately tested—under government-established regulations, as well as voluntary, industry-driven standards designed to reflect company's desire for quality.

When considering which of these industry-driven standards are the most appropriate for a given product, important consideration should be given to: ①What your product's eventual use will be. ②How it can be most effectively (and cost-effectively) manufactured. ③Whether or not its durability and lifespan will meet customer expectations.

As exemplified by the textile and clothing industry, understanding how the product should match the customer's specific needs will help the producer determine the best testing procedures to use. Even more important is to take into account the factors which the buyers might not necessarily have at the forefront of their minds while shopping. For instance, when people shop for leisure clothing, they generally think about such factors as comfort, style and cost. Durability might not heavily influence their decision at the time of purchase. However, if six months down the road the article begins to fray or lose its color, they will be very unlikely to choose the same brand again.

When testing is considered, the large variety of textile applications lead to a multiplication of tests with different experimental conditions in order to simulate as closely as possible the conditions the product must face in real life. This is known as "testing validity" and it is the litmus test for whether or not a particular quality test should be incorporated into the testing program.

It takes a lot of trial, error and further adjustments to previous testing standards before a test comes close to replicating real-world conditions. A pair of heavy construction pants must be subjected to tests that best replicate the daily battering from weather, mud and sustained contact with hard construction materials, formal wear will be measured more by its aesthetic qualities—the integrity of the stitching, or the quality of the dyes used and the process by which they are applied. One of the benefits of hiring an experienced Quality Assurance and Lab Testing Service for your textile products is that they can help you determine and incorporate testing validity into your product line from the start.

Aside from the tests performed by the manufacturer, a third-party certification may be sought for the finished product. This assessment is conducted for a fee by an independent laboratory. It warrantees that the product complies with a certain set of requirements referring to safety, quality, and/or performance. Tests are performed using specific standard test methods which were issued by one or more standardization bodies.

The objective of standardization is ensuring the consistency of essential features of goods and services (e. g. , quality, ecology, safety, economy, reliability, compatibility, interoperability, efficiency, and effectiveness) so that the product, process, or service is fit for its purpose. It can also allow a widespread adoption of formerly custom processes. The primary activity of standardization bodies is producing technical standards to address the needs of a relatively wide range of users: tasks consist in developing, coordinating, promulgating, revising, amending, reissuing, and interpreting documents that provide requirements, specifications, test methods, guidelines, and characteristics.

Standardization bodies exist at the international, regional, and national level. There are main standards for textiles and apparel testing including ISO (International Organization for Standardization), GB (China), EN (Europe), JIS (Japan), BS (UK), ASTM (North America), AATCC (North America). For instance, ISO is an international standard-setting body composed of representatives of national standards organizations from 164 countries, one per country. A technical committee on textiles, ISO/TC 38, was formed from the beginning to coin test methods, terminology, and definitions of fibers, yarns, threads, cords, rope, cloth, and other fabricated textile materials, and elaborate specifications for textile products. ISO/TC 38 also covers textile industry raw materials, auxiliaries, and chemical products required for textile processing and testing. A total of 382 standards are under its direct responsibility or that of its subcommittees. For another important internation standards body ASTM, Committee D13 on Textiles, consisting of over 500 members, and developer of over 300 standards, provides the technical direction for the program. Test information generated in the program is utilized by Committee D13 to determine if modifications to ASTM documents or new standards are warranted.

Generally, mandatory national and international compliance standards for textiles focus more on product safety and environmental protection, voluntary standards address aspects of quality, durability and style—the make-or-break factors in the eyes of the consumer.

In recent years, the quality of textiles has become the focus of consumer attention. With the increasing frequent spot checks on textiles by national regulatory agencies, textile and clothing quality issues have become important factors affecting textile companies' product access, brand reputation, and market position. If a producer wants to sell the textile products in China, the requirements of GB 18401—2010 should be met to control hazardous substances in textile products, improve textile product quality, ensure people's basic safety and health and to enforce the general safety and technical specifications. This mandatory national standard divides all textile products into three categories: Infant products, skin contact products and non-skin contact products. Products are assessed in comparison with the specific technical requirements listed in the below table (Table 47.1). Infant products should comply with the requirement of Type A, skin contact products should at least comply with Type B and non-skin contact should at least comply with Type C. All components that make up an infant product shall meet the Type A technical requirements as specified by the table below. The component less than 1% of the whole weight of non-infant products can be exempted from testing.

Table 47.1 General technical safety specifications for textile products

Items			Test standards	Type A	Type B	Type C
Formaldehyde content/(mg/kg) ≤			GB/T 2912.1	20	75	300
pH[a]			GB/T 7573	4.0-7.5	4.0-8.5	4.0-9.0
Dye colour fastness[b] (level) ≥	Water resistance	Color deterioration, staining	GB/T 5713	3-4	3	3
	Colorfastness to acidic perspiration	Color deterioration, staining	GB/T 3922	3-4	3	3
	Colorfastness to alkali perspiration	Color deterioration, staining	GB/T 3922	3-4	3	3
	Colorfastness to Coulomb friction	—	GB/T 3920	4	3	3
	Colorfastness to saliva	Color deterioration, staining	GB/T 18886	4	—	—
Odor			GB/T 18401 6.7	N/l		

Continued

Items	Test standards	Type A	Type B	Type C
Decomposable carcinogenic arylamine dyes[c] (mg/kg)	GB/T 17592 GB/T 23344	Forbidden		

a. The pH can be adjusted to 4.0 – 10.5 for non-end products that must undergo wet processing in further processing procedures.

b. Not applicable to: non-end products that must undergo washing and fading processes, products that are of original color or bleached products; traditional hand-colored products including tie-dyed or wax printed products. Colorfastness to saliva tested only for baby/kids products.

c. Please see appendix C for the list of decomposable carcinogenic arylamine dyes in GB 18401—2010, limit value: ≤20mg/kg.

With the rapid development of science and technology, new emerging applications are deepening at the same time promote more traditional textile products update. Smart textiles have seen tremendous growth over the past few years, especially in terms of research funding and development of start-ups. However, very few products have, so far, succeeded in reaching the consumer market. Indeed, one of the main challenges remains the lack of standardized test methods to evaluate the properties and performance of smart textiles on an objective and comparable basis. Although several testing methods are developed and used by various companies and research centers that produce such active materials, there is a growing need for organizations such as the ISO, ASTM, etc. to become involved in order to standardize test methods.

New Words and Expressions

stringent 严格的，严厉的
litmus 石蕊（试剂）
aesthetic 美的，美学的，审美的
stitching 针脚，缝合，缝纫
interoperability 互通性，互操作性
infant products 婴幼儿产品
formaldehyde content 甲醛含量
color fastness 染色牢度
water resistance 耐水洗
color change 变色
staining 沾色
acid sweat resistance 耐酸汗渍
alkaline sweat resistance 耐碱汗渍
dry rubbing resistance 耐干摩擦
saliva resistance 耐唾液
odor 异味

Notes to the Text

1. the distilled wisdom：浓缩的智慧，智慧的精华。
2. construction pants：建筑工裤。
3. the litmus test：试金石，立见分晓的检验办法. Litmus（石蕊）是有机物，它是一种弱的有机酸。石蕊试纸是常用的试纸，是一种检验溶液酸碱性的方法，有红色石蕊试纸和蓝色石蕊试纸两种。碱性溶液使红色试纸变蓝，酸性溶液使蓝色试纸变红。
4. aesthetic qualities：美学特征，审美特征。

5. product line：生产线。
6. standardization bodies：标准制定单位或组织。
7. ISO（International Organization for Standardization），GB（China），EN（Europe），JIS（Japan），BS（UK），ASTM（North America），AATCC（North America）：世界主要标准或标准组织：ISO 国际标准组织；GB 中国国家标准，我国强制性认证(CCC 认证)所依据的标准为 GB；EN(European Norm)欧洲标准，产品通过了 EN 的测试标准才能拿到 CE 认证；JIS 日本工业标准；BS(British Standard)英国标准；ASTM(American Society of Testing Materials）美国材料实验协会；AATCC（American Association of Textile Chemists and Colorists)美国纺织化学师与印染师协会。
8. GB 18401—2010 National General Safety Technical Code for Textile Products：国家纺织产品基本安全技术规范。
9. decomposable carcinogenic arylamine dyes：可分解致癌芳香胺染料。
10. The pH can be adjusted to 4.0–10.5 for non-end products that must undergo wet processing in further processing procedures：后续加工工艺中必须要经过湿处理的非最终产品，pH 可放宽至 4.0~10.5。
11. Not applicable to：non-end products that must undergo washing and fading processes，products that are of original color or bleached products；traditional hand-colored products including tie-dyed or wax printed products. Colorfastness to saliva tested only for baby/kids products：对需经洗涤褪色工艺的非最终产品、本色及漂白产品不要求；扎染、蜡染等传统的手工着色产品不要求。耐唾液色牢度仅考核婴幼儿纺织产品。

Questions to the Text

1. Why we need standards for textiles and apparel?
2. What is the purpose of standardization?
3. What is the difference between mandatory standards and voluntary standards ?
4. Discuss the specifications for textile products of GB 18401—2010 with your friends.

Lesson Forty-Eight

Textile Legislation *

语言微课堂

To help provide consumers with important information at the point of sale, and to provide protection to some degree, the U.S. government has enacted various rules and regulations that pertain to textile fabrics and products. These rules and regulations are designed to provide protection and sufficient information concerning various aspects of textiles so that consumers have the necessary data to provide long term care, to make knowledgeable selections, and/or to feel somewhat safe and secure regarding the purchase of the item.

Wool Products Labeling Act

The Wool Products Labeling Act (WPL) is a measure to protect producers, manufacturers, distributors, and consumers from the unrevealed presence of substitutes and mixtures in spun, woven, felted, or otherwise manufactured wool products.

The act requires that any textile product of wool or part wool must be labeled to indicate the quality and type of wool present. Several terms are defined by the act and its amendments. Since the enactment of the legislation there have been several amendments. Current definitions used in labeling wool products include the following.

Wool means the fiber from the fleece of the sheep or lamb or hair of the Angora or Cashmere goat that has never been reclaimed from any woven or felted wool product.

Recycled wool means ① the resulting fiber when wool has been woven or felted into a wool product which, without ever having been utilized in any way by the ultimate consumer, subsequently has been made into a fibrous state, or ② the resulting fiber when wool or reprocessed wool has been spun, after having been used in any way by the ultimate consumer, subsequently has been converted back into a fibrous state.

Wool product means any product, or any portion of a product, which contains, purports to contain, or in any way is represented as containing wool, reprocessed wool, or reused wool.

Reprocessed wool fits definition 1 for recycled wool, while reused wool fits definition 2 for recycled wool. The term recycled wool supersedes the use of reprocessed wool and reused wool as of the amendment of 1980. However, the terms reprocessed and reused still appear in some cross references concerning the WPL and the labeling of wool products.

Fibers from animals other than the sheep may be identified by the name of the animal as long as it is clear that it is a type of animal hair fiber. However, the use of special animal names is not required; thus, any fiber, such as cashmere, may be identified and labeled as wool without any other terminology.

The Federal Trade Commission is responsible for the enforcement of the WPL and provides both manufacturers and consumers with helpful information concerning this act.

Other terms that may be used in identifying wool include the following. Virgin or new may be used with the term wool only when the fiber is new and has never been reclaimed from any type of woolen product. The term fur fiber may be used to identify fibers taken from animals normally used for fur. The term may not be used with wool or specialty fibers. The use of the name of one of the specialty animals, such as the cashmere goat. Angora goat (mohair), vicuna and others may be applied only when fiber from those animals is actually involved. The term lamb's wool is not defined in the rule as it now stands. However, the term is used to represent wool taken from animals under eight months of age.

In general the quality of recycled wool is usually lower than that of new fibers. During the garneting process that separates the fibers back into a fibrous mass from yarns or fabrics, some damage may occur to the fibers. However, the warmth factor is not affected, and recycled wool is satisfactory in such products as interlinings, padding for carpeting, inexpensive blankets, and similar products.

Textile Fiber Products Identification Acts (TFPIA)

The TFPIA has been amended several times since its birth. The most important amendments have been the addition of generic terms to be used in labeling textile fibers. The act requires that textile products be labeled to identify fiber content by percentage (except for fibers percent in amounts less than 5%, which may be identified as "other"). The act further established generic terms for manufactured fibers.

The act further indicates that natural fibers, other than wool, which is included under the Wool Products Labeling Act, must be identified by name. All fibers must be identified by percentage, with the fiber of highest percentage given first, followed by the next percentage component and so forth, ending with the fiber with the lowest percentage. The act exempts some textile products such as certain coated fabrics, industrial fabrics of some types, and upholstery that is already installed on furniture. Both the textile product itself and any advertising of such products must identify the fiber content.

The TFPIA identified generic terms that must be used for all manufactured fibers. If trade names or trademarks are to be used, they must be accompanied by the generic term for the fiber. The trade name or trademark shall be capitalized; the generic term is not, except where both words are completely written in capital letters.

The act requires that correct fiber content be transmitted by fiber producers to product manufacturers, who, in turn, must provide adequate and accurate information concerning fiber content to the retailer. The retailer is responsible for labeling the product at the consumer level.

When fabrics are composed of two or more fibers, the label must identify the actual percentage of each fiber in descending order as noted above. If less than 5% of a fiber is present, this low percentage may be indicated as "other fibers"; however, if the fiber serves a clearly established purpose or function, such as an elastomeric fiber, it may be identified by percentage, generic name, and significant property imparted.

Clear and concise guidelines concerning the TFPIA have been prepared by the FTC(Federal Trade Committee), and both retailers and consumers may obtain copies from the FTC free of charge. The FTC is responsible for the enforcement of the act. However, the real responsibility for enforcement may, in fact, rest with the consumer. Unless consumers are willing to report evidence of noncompliance with the act, the actual value of the legislation may be reduced owing to lack of enforcement funds.

The laws and regulations related to textiles and apparels in China

The laws and regulations related to textile and apparel products in China include Product Quality Law of the People's Republic of China, Law of the PRC on the Protection of the Rights and Interests of Consumers, Law of the People's Republic of China on Import and Export Commodity Inspection and It's Implementation Regulations, Standardization Law of the People's Republic of China and It's Implementation Regulations.

Except for mandatory standards, textile products should be in accordance with their stated standard while to enter the China market. According to the People's Republic of China Standardization Law, China's current standards are divided into mandatory or recommended standards. The recommended standard is mostly technical standard, and the mandatory standard is of regulatory nature. Some Chinese mandatory textile standards are listed in Table 48.1.

Table 48.1 Examples of some of the main Chinese mandatory textile standards

Mandatory standard code	Standard name	Applicable for
GB 18401—2010	National general safety technical code for textile products	The textile products used for clothing, decoration and household which are produced and sold within the territory of China
GB 31701—2015	Safety technical code for infant and children's textile products	Infants and children textile products sold in the Chinese domestic market
GB 20400—2006	Leather and fur—Limit of harmful matter	Daily used leather and fur products
GB 21550—2008	Restriction of hazardous materials in polyvinyl chloride artificial leather	Polyvinyl chloride artificial leather
GB 18383—2007	General technical requirements for products with filling materials	Filling materials

New Words and Expressions

legislation 法规
enact 颁布
pertain 关于
Wool Products Labeling Act(WPL) 《羊毛产品标签法》
distributor 批发商
amendment 修改
purport 似乎
angora goat 安哥拉山羊
cashmere goat 山羊
recycled wool 再生毛
enforcement 实施，执行
interlining 中间衬料
Textile Fiber Products Identification Acts (TFPIA) 《美国纺织纤维产品鉴定条例》
generic (生物的)属
exempt 免除
FTC(Federal Trade Committee) 美国联邦贸易委员会
filling materials 填充物

Notes to the Text

1. unrevealed presence of substitutes：隐藏的替代物，掺假。
2. to indicate the quality and type of wool present："present"意思是"存在的"。
3. that has never been reclaimed from any woven or felted wool product："reclaimed"意思是"回收"。
4. reprocessed wool：经过再次加工的毛。
5. reused wool：再次利用的毛。
6. The term recycled wool supersedes the use of reprocessed wool and reused wool as of the amendment of 1980：在 1980 年的修正案中用"再生毛"取代了"再次加工的毛和再次利用的毛"。
7. cross reference：交叉参考。
8. purports to contain：声称含有。
9. specialty animal：特种动物(英国拼法：speciality animal)。
10. The term lamb's wool is not defined in the rule as it now stands：羔羊毛的定义与其目前表示的含义有区别。
11. The act further indicates that natural fibers, other than wool：该法令进一步指出，除羊毛外的天然纤维。
12. The TFPIA identified generic terms that must be used for all manufactured fibers：所有人造纤维必须使用纺织纤维产品鉴别法确定的名称。
13. However, the real responsibility for enforcement may, in fact, rest with the consumer："rest with the consumer"意思是"取决于消费者"。
14. Product Quality Law of the People's Republic of China, Law of the PRC on the Protection of the Rights and Interests of Consumers, Law of the People's Republic of China on Import and Export Commodity Inspection and It's Implementation Regulations,

Standardization Law of the People's Republic of China and It's Implementation Regulations：指中华人民共和国产品质量法、中华人民共和国消费者权益保护法、中华人民共和国进出口商品检验法、中华人民共和国标准化法及其实施条例。

Questions to the Text

1. Explain the meaning of recycled wool.
2. How to show the fiber content in label?
3. What is wool according to WPL?
4. What is wool product?
5. Give examples of specialty fibers.

Lesson Forty-Nine

High Performance Fibers *

语言微课堂

Glass fibers

There are two processes for manufacturing glass yarns, these are for continuous filament and staple fiber. The raw materials are mixed in accordance with formulas which are adapted to the end use of the fiber. These materials are drawn first in the batch bins and into the batch mixed and then into the batch melting glass refining tank. In the continuous filament process, the fibers are drawn mechanically, while molten, through temperature resistant alloy metal feeders having more than 100 small orifices. As fine strands of molten glass emerge through these holes, they are gathered together, run over a pad where a sizing is applied and then carried to a high speed winder which draws the filament at a rate of more than a mile a minute.

In the staple fiber process, the staple fiber is formed having a long staple characteristic, employing jets of compressed air to make the molten glass into fine fibers. The molten glass flows through orifices and temperature resistant bushings at the base of each furnace. The impact of the compressed air yanks the thin strings of molten yarn into fine fibers varying in lengths of 20–30 centimeters and drives them down into a bobbin drum on which they form a thin veil resembling a cobweb.

As the fibers hit the drum, instantaneously the fibers pass through a spray of lubricant and a drying flame. This web of fiber is then gathered from the drum into a sliver that is slightly drafted in the winding operation so that the majority of the fibers lie parallel with the length of the sliver. These slivers can be further reduced in diameter and then twisted and plied in yarns of various sizes using textile machinery and processes similar to those employed for other staple materials. Currently, staple fibers are used for architecture, vehicles, shipping, space ships, refrigerators, etc. , as sound-, heat-insulating materials and heat reserving materials. By contrast, because filament yarn can be woven, filament yarns are used for electric insulating material or flame retardant curtains as well as plastic reinforcement materials as staple fibers.

Features

1. High tensile strength, excellent resistance to heat, and flame retardancy.
2. Excellent heat-and electric-insulating.
3. No water absorption.
4. Very resistant to acid and other chemical agents. Non-attackable by molds and insects.

Carbon fibers

Carbon or graphite fibers have been developed for use in various types of industrial textiles and in some types of consumer goods. Carbon fibers are divided into two categories. One is PAN (polyacrylonitrile)-type carbon fibers manufactured by burning a special acrylic fiber. Another is pitch-type carbon fibers manufactured from pitch (remaining products of distilled petroleum or coals). These fibers are made by converting precursor filament into 95 to 99% carbon.

Precursor filaments such as acrylic or pitch are carbonized at high temperature. Graphite fibers are formed under treatment at temperatures above 2,500℃, which converts the precursor filament to 99% carbon. Fibers treated at temperatures up to 1,500℃, which are converted to 95% carbon, are those identified as carbon fibers. Carbon fibers have an extremely high modulus and a low elongation; thus they are highly suited for uses where high strength and little or no stretch are desired. In addition, they are highly resistant to heat and chemical agents.

The tenacity of the fibers varies from about 3 to more than 19 grams per denier; elongation is less than 1%. Tensile strength of fabrics is higher than any other commercially available fiber except for sapphire whiskers and graphite in whisker form. Fiber density is approximately 1.5 grams per cubic centimeter. A few carbon fibers may have a density as high as 2, but these are usually designated for very special uses.

Carbon fibers have a diversity of applications. They are used to reinforce resins and metals to provide structural materials with high strength and stiffness and light weight. The resulting composites are used in aerospace applications, rotor blades for helicopters, sporting equipment such as golf club shafts, bicycle parts, skis, and machinery parts. The carbon filaments are also used to make fabrics that are flameproof and can be used for a variety of protective apparel; they provide for the military where nuclear, biological and chemical resistance is important; and they are being promoted as an alternative to asbestos.

Features

1. High tensile strength and little extension even by pulling fibers.
2. Excellent in resistance to abrasion and dimensional stability.
3. Excellent in resistance to heat and to chemical agents.
4. Excellent in electrical conductivity.

专业微课堂：
碳纤维

Aramid fibers

Aramid fibers are divided into para-and metha-types depending on their components. A para-type aramid fibers are superior in tensile strength and elasticity. A metha-type aramid fibers are superior in heat resistance and flame retardancy, which are used for clothing of fire-fighters, car racing suits, and uniform for high-temperature workings, reflecting its excellent fire retardancy.

The most popular para-type aramid fiber on the world market is Kevlar. Kevlar aramid fiber has an extremely high tenacity, approximately 22 grams per denier, which is more than five times the strength of a steel wire of the same weight and more than twice the strength of high-tenacity industrial nylon, polyester, or fiberglass. It has an unusually high initial modulus of 476 grams per denier. This is about two times greater than fiberglass or steel, four times greater than industrial polyester, and nine times greater than industrial nylon. Moisture regain is about 7%. Elongation is low, only about 4%.

Thermal properties of Kevlar are superior. It has outstanding stability to heat, retaining a high percentage of strength after exposure to temperatures to 260℃.

A popular use of Kevlar aramid fibers is in the manufacture of tire cord. The advantages of aramid-reinforced tires are that they are lighter in weight than those using glass or steel fiber, they give good performance at sustained high speeds, and they have a lower resistance to running and therefore save on fuel consumption. They run cooler than other tires. They are made on existing equipment so cost less than steel-reinforced tires, and they have outstanding durability.

Kevlar is used in a variety of applications. One of the most interesting and important uses is in body armor or bulletproof vests, and it is credited with saving numerous lives of police officers. This use depends on the high strength of the fiber. Other uses for the fiber include: bracing belts, high-pressure hoses, conveyor belts, rope and cables used as antennae supports, oceanographic cables, crane ropes, boat rigging, helicopter hoist cables, special cables for deep-sea workstations, reinforcement for concrete in various types of constructions, many parts of the space shuttles, space vehicles, body panels for race cars, boats of various types.

Words and Expressions

antenna　(antennae) 天线
hose　软管，水龙带
string　线，细绳
instantaneous　瞬间的，即刻的，即时的
glass fiber　玻璃纤维
formula　配方，处方
bin　箱，池，仓
feeder　喷丝头
orifice　外孔，小孔
pad　衬垫
sizing　上浆，浆纱
bushing　套筒，套管
yank　猛拉
veil　面纱
cobweb　蜘蛛网
drum　筒管
ply　合股
carbon fiber　碳纤维
graphite fiber　石墨碳纤维
industrial textiles　产业用纺织品
pitch　场地，这里指沥青
precursor　母体，前驱体
modulus　模量
tenacity　强度
sapphire whisker　蓝宝石晶须
resin　树脂
structural materials　结构材料
stiffness　刚度
golf club shaft　高尔夫球杆
ski　滑雪板
asbestos　石棉
aramid fiber　芳纶
para-and metha-types　对位与间位
Kevlar　凯夫拉芳纶

industrial nylon 工业用尼龙
initial modulus 初始模量
thermal properties 热性能
durability 耐久性
space vehicle 太空车，宇宙飞船
body armor 护身盔甲
bulletproof vest 防弹背心
bracing belts 船用索带
deep sea workstation 深海工作站
oceanographic cable 深海缆绳
boat rigging 船缆，船索
space shuttle 航天飞机

Notes to the Text

1. in accordance with formulas：依照配方。
2. batch bin：料仓。
3. while molten, through…：中的“while molten”为插入语，表示状态。
4. the staple fiber is formed having a long staple characteristic：形成的短纤维具有加长短纤维的特点(玻璃短纤维的长度一般均远远大于常规短纤维)。
5. as the fibers hit the drum：“hit”在这里表示短暂的接触。
6. lie parallel with the length of the silver：沿纱条的长度方向平行排列。
7. similar to those employed for other staple materials：在句中作“textile machinery and processes”的定语。
8. Carbon or graphite fibers：碳纤维或石墨碳纤维。这是为了区别纤维含碳量而使用的两个名称，在一般情况下，可通称为碳纤维。
9. Tensile strength of fabrics is higher than for any other commercially available fiber except for sapphire whiskers and graphite in whisker form：(用碳纤维生产的)织物的强度，除了蓝宝石晶须和石墨晶须(织物)外，比其他任何商业用纤维的织物强度都高。
10. a diversity of applications：广泛的用途，各种各样的用途。
11. flameproof：防火。某些单词与“proof”组合，可以构成一个新的单词，表示“防……”的意思。在纺织行业中常用的例子有：waterproof（防水），oil-proof（防油），downproof（防羽绒），heat-proof（防热）等。
12. and they are being promoted as an alternative to asbestos：并且它们正在被推荐作为石棉的替代物。石棉作为传统的保温隔热材料，由于具有致癌作用，已在许多国家的某些行业被禁止使用。
13. Kevlar：凯夫拉芳纶，是美国杜邦公司生产的芳纶的商品名称。
14. fiberglass：玻璃纤维，与“glassfiber”意义相同。
15. more than twice the strength of high-tenacity industrial nylon, polyester, or fiberglass：在“fiberglass”后面省略了“of the same weight”。
16. tire：轮胎(通常指外胎)，这是美国拼法，英国拼法为 tyre。
17. aramid-reinforced tires：用芳纶增强的轮胎。
18. at sustained high speed：在连续高速行驶时。
19. lower resistance to running：对行驶的阻力小。
20. They run cooler than other tires：与其他轮胎相比，它们不易升温。

21. existing equipment：现有设备，指不使用特殊加工设备。
22. it is credited with：由于……而获得声望、受到赞扬。
23. antennae supports：天线的固定。“antennae”是“antenna”（天线）的复数形式。
24. reinforcement for concrete：对水泥材料的增强。
25. space vehicles：宇宙飞船。
26. space shuttle：专指航天飞机。

Questions to the Text

1. How to make the continuous glass filaments?
2. How to make the staple glass fibers?
3. What are the major differences between the processes of making glass filament and glass staple fiber?
4. Why the carbon fiber has two names (the carbon fiber and the graphite fiber)?
5. What are the main properties of the carbon fiber?
6. Cite several uses for carbon fibers and indicate why they are good for some end uses?
7. What are the major properties of the aramid fibers?
8. The aramid fibers are generally found in what types of end-use products? Why?
9. Give names of some aramid products you know.

Lesson Fifty

Nanofibers*

With the development of nanotechnology during the past two decades, there have been a significantly increasing number of studies on nanofibers and their applications. The International Standards Organization (ISO) considers nanomaterials to be materials that are typically but not exclusively below 100nm in at least one dimension. However, in informal nonwovens, textile, and other engineered fibers industries, it has been well accepted that nanofibers are fibers with diameters smaller than 1,000nm. Nanofibers have emerged as exciting one-dimensional (1D) nanomaterials for a wide range of research and commercial applications due to the unique properties that arise from their size-dependent structural behavior, such as a diameter 1,000 times smaller than that of human hair, high surface area, large surface-to-volume ratio, large fiber length-to-diameter aspect ratio, excellent pore interconnectivity, 3D topography, flexible surface functionalities, and reasonable mechanical properties (e.g., stiffness and tensile strength).

There have been many techniques developed for the fabrication of nanofibers, such as electrospinning, self-assembly, phase separation, interfacial polymerization, template or pattern-assisted growth, vapor-liquid-solid growth and hydrothermal synthesis, etc.

Electrospinning

Electrospinning is currently the most promising technique to produce continuous nanofibers on a large scale because of its simplicity, low-cost set-up, and the possibility to control the nanofiber composition, diameter, and orientation in the function of the intended application. Electrospinning is a technique using electrostatic forces to fabricate nanofibers. The basic setup of the electrospinning method includes three major components: reservoir (including a syringe, syringe pump, and syringe needle), collector (rotating or stationary), and voltage supplier, as shown in Fig.50.1. During electrospinning, the liquid is extruded from the spinneret to produce a pendant droplet as a result of surface tension. Upon electrification, the electrostatic repulsion among the surface charges that feature the same sign deforms the droplet into a Taylor cone, from which a charged jet is ejected. The jet initially extends in a straight line and then undergoes vigorous whipping motions because of bending instabilities. As the jet is stretched into finer diameters, it solidifies quickly, leading to the deposition of solid fiber(s) on the grounded collector. In general, the electrospinning process can be divided into four consecutive steps: ①charging of the liquid droplet and formation of Taylor cone or cone-shaped jet; ②extension of the charged jet along a straight line; ③thinning of the jet in the presence of an electric field and growth of electrical bending instability (also known as whipping instability); and ④ solidification and collection of the jet as solid fiber(s) on a grounded collector. In many cases, randomly oriented fibers are deposited on a flat collector plate forming a non-woven mat of fibers. Controlled fiber deposition techniques are also applied for the fabrication of aligned nanofibers, on a rotation drum. However, electrospinning has limitations of low productivity, as solutions are usually fed at a low rate so as to produce fibers of low diameter. A variety of fascinating structures have been successfully prepared via electrospinning, such as lotus leaf, butterfly wings, helical, necklace-like, rice-grain shape, water strider legs, spider-web-like and tree-like structure. These structures have unique surface or inner multilevel structure, either micro-, nano-or micro-/nano-combined structures, and result in some advantages or excellent performance in the applications of filtration, self-cleaning materials, tissue engineering, sensors, catalysts, etc.

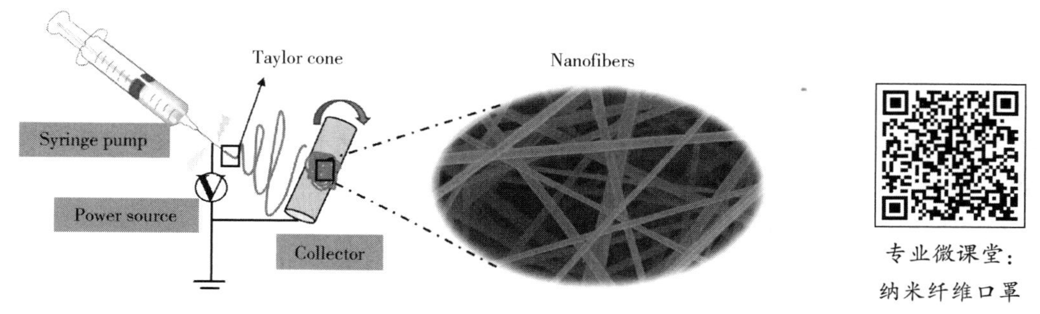

Fig.50.1 Schematic diagram of the electrospinning apparatus

Self-assembly

Self-assembly is a bottom-up process in which small molecules spontaneously assemble into well-ordered nanofibers. The formation of this structure is induced by many interactions, including chiral dipole/dipole interactions, π-π stacking, hydrogen bonds, non-specific van der Waals interactions, hydrophobic forces, electrostatic interactions, and repulsive steric forces. It produces nanofibers in high yield with low polydispersity, enabling further exploration of this method for developing "smart" biomaterial scaffolds for effective tissue regeneration.

Phase separation

Thermally induced phase separation was commonly employed during the early days to produce porous polymeric scaffolds. The method was explored further to produce nanofibrous 3D structures from a variety of biodegradable polymers. There are five basic steps in this technique: ①polymer dissolution, ② phase separation and gelation, ③ solvent extraction from the gel with water, ④freezing, and ⑤ freeze-drying under vacuum. The selection of proper solvent is considered as one of the most critical step of nanofibrous structure formation during this process. The formation of the nanofibrous structure is postulated to be caused by spinodal liquid-liquid phase separation of the polymer solutions and consequential crystallization of the polymer rich phase. The method does not require specialized instruments and it also allows for batch to batch consistency, while the architecture and scaffold properties can be controlled easily by varying the polymer concentration, gelation temperature/time, solvent, and freezing temperature. Macroporosity was another feature that could be obtained within these scaffolds by incorporating porogens such as salt or sugars into the polymer solution during the phase separation process. Such 3D microporous structures are advantageous to the cells to absorb nutrients, receive signals, and to discard wastes. The presence of both nano-and macro-structures at the nanofiber level provide additional benefits to cell distribution and response.

Interfacial polymerization

Interfacial polymerization was developed as a template free approach to synthesize large quantities of pure, uniform polyaniline nanofibers. Typically, an immiscible system consisting of an organic solution of aniline and an aqueous solution of ammonium peroxysulfate and acid is left to stand for some time. Nanofibers are obtained exclusively because the green polyaniline nanofibers formed in the early stages of aniline polymerization at the organic/aqueous interface diffuse into the aqueous phase, thus preventing secondary overgrowth of polyaniline on the nanofibers that produce agglomerates normally observed in traditional chemical oxidative polymerization of aniline in homogeneous aqueous systems. The nanofibers formed are typically twisted and interconnected with diameters ranging from 30 to 50nm and length about 500nm to several micrometers. Nanofiber diameters are affected by the choice of acid used, while quality and uniformity of the nanofibers are controlled by acid concentration; the higher the acid concentration, the higher the fraction of nanofibers in the final product.

Template or pattern-assisted growth

This method facilitates the production of controlled arrangements of nanofibers. Porous template-based growth of nanofiber arrays can be obtained using well established techniques such as electrochemical deposition and template filling. Various inorganic materials such as metal oxides, metals, semiconductors, and organic materials have been synthesized as nanofiber arrays using this technique. The main drawback of this method can be listed as degradation of the template under longer polarizations and non-uniform pore filling for high-aspect-ratio nanostructures. Mesoporous metal oxides which are widely used for sensing and energy applications can be produced with well-defined and ordered porous structure by means of using surfactant as templates through sol-gel method. In this method, desired pore size, nanostructure morphology, size distribution, and density of pores can be obtained by selecting the appropriate template.

Vapor-liquid-solid (VLS) growth

This method is well known for synthesizing defect free 1D nanostructures for a wide range of materials. The parameters of the nanofiber such as diameter, length, and composition as well as growth direction can be effectively controlled by understanding the mechanism of the VLS technique. The formation of metal nanodroplets from gaseous precursors plays a major role in the growth of nanofibers using this method. At first the dispersed metallic nanocrystals on a single crystalline substrate is melted in a tube furnace. Various process gases introduced during this process lead to the saturation of the molten metal nanodroplets which act as catalysts resulting in the continuous precipitation of single crystalline nanofibers thereby promoting the unidirectional growth. Both hybrid and doped nanofibers can be produced using this method. Growth orientation in particular planes can be effectively controlled by appropriate substrate selection and optimizing the corresponding temperature and pressure during growth. The diameter of the 1D nanostructure produced in this method is controlled by altering the growth parameters by tuning the properties of the liquid alloy droplet. The preparation of the nano-sized droplets on the substrate plays a major role in 1D nanostructure growth since it determines the kinetics of supersaturation and nucleation occurring at the liquid/solid interface resulting in the axial crystal growth. Metal catalysts activate the sites where the nanofibers are to be grown and hence it determines the position of the 1D nanostructures. Pressure of the source species also forms an integral role in growth rate which is directly proportional to the whisker diameter and hence the 1D nanostructures grows faster axially in whiskers of larger diameter.

Hydrothermal synthesis

This method involves the growth of the nanofibers in a heated liquid solution under pressure of 101.325kPa(1atm) in an autoclave at a temperature of about 100–300℃. The growth phase is dominated by the chemical decomposition during which the thermally degraded reactive ions from the precursors in the solution contribute to the growth of the nanofibers. The growth can be facilitated in a particular orientation by using appropriate catalysts which serves to lower the surface energy. This method was first employed for growing single crystals of semiconductors which was later improvised and modified for developing other nanostructures. The growth of the nanofibers can be controlled by manipulating the parameters such as temperature, precursor concentration, and pH, etc. The main disadvantage of this method is the low levels of crystallinity of the grown nanofibers. In hydrothermal synthesis, the properties of the 1D nanostructures are defined by the process kinetics which in turn dependent on the parameters such as temperature and pressure in the system, duration of the synthesis, and the initial pH of the solution medium. Salts called mineralizers also play a major role in providing supercritical conditions for hydrothermal synthesis since it forms the hydrothermal solution which determines the solubility of the metal oxides to be processed. Porosity of the synthesized 1D nanostructure can be controlled by selecting suitable surfactants for the targeted nanomaterial. The morphology of the metal oxide nanostructures is dependent on the surface-active agents (SAA) which has immense influence on the hydrothermal growth of oxide compounds.

New Words and Expressions

topography 地形
self-assembly 自组装
phase separation 相分离
interfacial polymerization 界面聚合
reservoir 储液器，储液装置
stationary 固定的
spinneret 喷丝头
pendant 悬垂的
electrification 带电；充电
whip 鞭子；鞭打
consecutive 连贯的
lotus leaf 荷叶
butterfly wings 蝴蝶翅膀
helical 螺旋状的
necklace-like 项链状
rice-grain 稻粒
water strider legs 水黾腿
spider-web-like 蛛网状
tree-like 树状的
spontaneously 自发地，自然地；不由自主地
chiral 手性的
dipole 偶极
steric 空间的，立体的
polydispersity 多分散性
gelation 凝胶化
postulated 被假设的，假定
spinodal 旋节线；拐点
macroporosity 大孔隙度；宏观孔隙度
porogen 致孔剂
nutrient 营养物，滋养物
polyaniline 聚苯胺
immiscible 互不相溶的
ammonium peroxysulfate 过硫酸铵
surfactant 表面活性剂
nanodroplet 纳米液滴
nanocrystal 纳米晶体
tube furnace 管式炉
saturation 饱和
precipitation 沉淀
unidirectional 单向的
alloy 合金
nucleation 成核现象；集结
integral 完整的，整体的；构成整体所必需的
whisker 晶须；胡须，腮须
autoclave 高压灭菌器；高压锅
manipulate 操纵，操作；巧妙地处理
mineralizer 矿化剂，造矿元素
targeted 定向的，被定为攻击目标的
immense 巨大的，广大的

Notes to the Text

1. the possibility to control the nanofiber composition, diameter, and orientation in the function of the intended application：根据预期应用的功能控制纳米纤维的组成、直径和方向的可能性。
2. Upon electrification, the electrostatic repulsion among the surface charges that feature the same sign deforms the droplet into a Taylor cone, from which a charged jet is ejected：带电时，具有相同符号的表面电荷之间的静电斥力使液滴变形为泰勒锥，从中喷射出带电的射流。
3. whipping instability：鞭动不稳定。

4. it also allows for batch to batch consistency：可以实现批次间的一致性。

Questions to the Text

1. What are the advantages of nanofibers?
2. What are the preparation methods of nanofibers?
3. Describe the electrospinning process.

Lesson Fifty-One

Overview of Industrial Textiles

语言微课堂

Industrial textiles are specially designed and engineered structures that are used in products, processes or services of mostly non-textile industries. According to this definition, an industrial textile product can be used in three different ways:

(1) An industrial textile can be a component part of another product and directly contribute to the strength, performance and other properties of that product, e. g. , tire cord fabric in tires.

(2) An industrial textile can be used as a tool in a process to manufacture another product, e. g. , filtration textiles in food production, and paper machine clothing in paper manufacturing.

(3) An industrial product can be used alone to perform one or several specific functions, e. g. , coated fabrics to cover stadiums.

Another indication of the definition above is that unlike ordinary textiles which have traditionally been used by the consumer for clothing and furnishing, industrial textiles are generally used by professionals from industries of non-textile character in various high-performance or heavy duty applications.

The term "industrial textiles" is the most widely used term for non-traditional textiles. Other terms used are "technical textiles" "high performance textiles" "high tech textiles" "engineered textiles" "industrial fabrics" and "technical fabrics". Industrial textiles are not in the spotlight like flashy brand-name apparel textiles and live almost a secret life, often hidden from the eye of the end user. And the manufacturers of high-technology industrial textiles are very secretive about their manufacturing processes and the design of their fabrics.

Although the beginning of industrial textiles may be as old as traditional textiles dating back to several thousand years ago, industrial textiles are considered to be a little "younger" than traditional textiles. The history of modern industrial textiles probably began with the canvas cloth used to sail ships from the old world to the new across the ocean. Later, hemp canvas was used on covered wagons to protect families and their possessions across the land. Fabrics were used in early cars as "rag-tops" to keep out the weather and as seat cushions for passenger comfort. Fabrics offered the advantage of light weight and strength for early flying crafts in the air. The wings of the earliest airplanes were made of fabrics. Industrial textiles are still used in hot air balloons and dirigibles.

The invention of man-made fibers in the first half of the 20th century changed the industrial textiles market forever. The first truly man-made fiber, nylon, was introduced in 1939. They offered high strength, elasticity, uniformity, chemical resistance, flame resistance and abrasion resistance among other things. New fabrication techniques also contributed to the improved performance and service life of industrial textiles.

By development of exceptionally strong high performance fibers in the 1950s and 1960s, the application areas of industrial fibers and fabrics were widened. Man-made fibers not only replaced the natural fibers in many applications, but also opened up completely new application areas for industrial textiles. For example, industrial textiles have played a critical role in space exploration. Spacesuits (Fig.51.1) are made of a layered fabric system to provide protection and comfort for the astronauts. Engineered textiles provided strong and lightweight materials for the lunar landing module and for the parachutes used to return the astronauts to earth in 1969. Military applications, especially during the global conflicts, expedited the development of technical textiles to better protect the soldiers. Today technical textiles are used extensively in military equipment and protective structures.

思政微课堂：
产业用纺织品成为
纺织行业新增长点

Fig.51.1 Space suit

Industrial textiles have been entering every aspect of human life. Thanks to advanced medical technology, today minute bundles of fibers are implanted in human bodies to replace or reinforce parts of the human body. Specially engineered textiles are used in airplanes, under highways, in transportation, and for environmental protection to name a few.

Industrial textiles make a vital contribution to the performance and success of products that are used in non-textile industries. For example, 75% of the strength of an automobile tire comes from the tire cord fabric used in the tire. Pure carbon fibers that are used in textile structural composite parts for aerospace, civil and mechanical engineering applications are on average four times lighter and five times stronger than steel.

Some of the modern industries simply would not be the same without industrial textiles. For example, the U. S. Department of Defense has in its inventory some 10, 000 items which are made entirely or partially from industrial textiles. The artificial kidney used in dialysis is made of 7, 000 hollow fibers and has a diameter of only two inches. Heat shields on space vehicles are made of textile fibers that can withstand several thousand degrees Fahrenheit. These are just some of the many examples which show the significance of industrial textiles in the journey of mankind.

Research and development work is being done continuously in the industrial textiles area. Both basic and application oriented research is conducted to find new materials, to improve the properties of the existing materials and products and to find new application areas for industrial textiles.

High-tech research is continuously being done at various levels of industrial textiles: polymers, fibers (the share of natural fibers is not expected to grow in industrial fabrics markets in the future), yarns, manufacturing methods, finishing and coating. Applied research has been proven very valuable to solve major industry problems.

The rapid progress in industrial textiles will increase the demand for highly technically skilled people for both production and research and development. The technical textiles industry offers careers in fiber chemistry, fiber, yarn and fabric production, product design and quality management in several engineering disciplines including mechanical, civil, chemical and materials engineering.

New Words and Expressions

industrial textiles 产业用纺织品
tire cord fabric 轮胎帘布
filtration 过滤
canvas 帆布
rag-tops 抹布
dirigible 飞船
lunar landing module 月球着陆模块
parachutes 降落伞
expedit 加快，加急
aerospace 航空航天
civil and mechanical engineering 土木和机械工程
artificial kidney 人工肾
dialysis 透析
Fahrenheit 华氏温度

Notes to the Text

1. Industrial textiles are not in the spotlight like flashy brand-name apparel textiles: 产业用纺织品不像华丽的名牌服装纺织品那样受到关注。
2. minute bundles of fibers: 是指细小的纤维束。

Questions to the Text

1. Please elaborate the application ways of industrial textiles.
2. What is the definition of industrial textiles?
3. What techniques should be improved to widen the application fields of industrial textiles?

Lesson Fifty-Two

Classification of Industrial Textiles*

Classification of industrial textiles is a challenging task. Industrial textile is a very diverse market segment of the textile industry. There are some products that can be used in many different applications. Each classification has advantages and disadvantages which can be a lengthy discussion. Classically, the classification of industrial textiles is made mainly on the basis of the final application.

Textile composites

A textile composite is made of a textile reinforcement structure and a matrix material. Textile reinforcement structures can be made of fibers, yarns or fabrics (woven, braided, knit and nonwovens) and are generally flexible. These structures are called textile preforms [Fig. 52.1 (a)]. Textile preforms can be in various shapes and forms. Matrix materials can be thermoplastic or thermoset polymers, ceramic or metal. The consolidation of the textile structure with the matrix material produces textile structural composites [Fig.52.1(b)].

The 21st century has already seen rapid advances in the field of textile-based composites. In addition to traditional 2D woven fabrics, multiaxial fabrics are now very widely used, with a rapidly growing number of applications for braided composites. Developments in 3D weaving, combining through-thickness reinforcement with excellent in-plane properties, provide new design solutions for situations where delamination must be avoided. Textile reinforcement is thus providing major new areas of opportunity for composite materials worldwide.

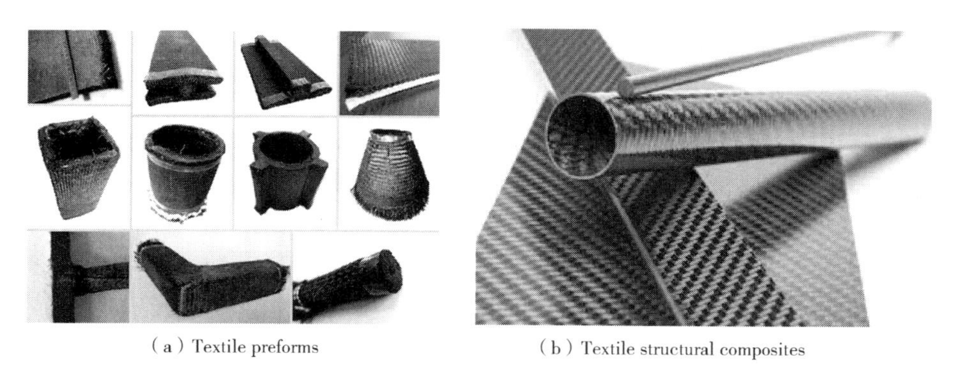

(a) Textile preforms　　(b) Textile structural composites

Fig.52.1 Textile preforms and textile structural composites

Filters

Textile materials especially woven and nonwoven fabrics are particularly suitable for filtration because of their complicated structures and considerable thickness. Textile fabrics are in fact a 3D network of fibers enclosing small pockets of void volume. Dust particles have to follow a "tortuous" path around textile fibers. Due to their structures, textile fabrics have high filtration efficiencies. The dust collection efficiency of fabric filters can range from 25 to 99.9% depending on the fabric type. High dust collection efficiencies and reasonable filter life (before plugging) can be obtained with woven and nonwoven fabrics (Fig.52.2). Textile materials do not restrict the flow of the fluid too much, yet they efficiently stop the particles.

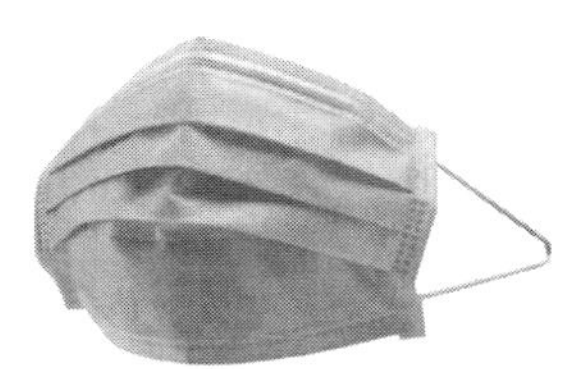

Fig.52.2 Nonwoven mask

思政微课堂：
中国防护服 守护世界

Geotextiles

Geotextiles can be separated into two terms, "geo" and "textiles". The word "geo" comes from the Greek meaning "earth", so geotextiles can be defined as the permeable textile materials that are used in combination with soil or any other civil engineering material (Fig.52.3). They are the largest group of geosynthetics in terms of volume. They are used in geotechnical engineering, heavy construction, building and pavement construction, hydrogeology, and environmental engineering. Geotextiles are traditional textile products such as woven and nonwoven fabrics. Knit fabrics are hardly used as geotextile materials. Recently, special structures were developed with the same applications as geotextiles. These new types of materials are generally very coarse compared to conventional-type geotextiles. Examples of these new structures are webbings, mats and nets.

The most important step toward the selection of geotextiles should be based upon "design by function route". This step can be successfully achieved with multidisciplinary approach based upon the combined efforts of various disciplines including textile, civil, chemical, and materials science.

Fig.52.3 Application of geotextiles in subgrade reinforcement

Medical textiles

Industrial textiles used in medical devices, surgeries, and other medical applications must meet stringent medical cleanliness and cytotoxicity standards to ensure they are safe for contact with patients. These fabrics must often exhibit a range of antibacterial and chemical resistance properties. In addition to protective medical apparel, textiles in fiber and fabric form are used for implants, blood filters, and surgical dressings. Woven and knitted materials in both synthetic and natural form play a part in the biotextile field, but nonwoven materials also have been proven to be effective and cost-efficient to be utilized in every area of medical and surgical textiles. Since there is such a broad range of properties in textile materials, the required properties of a specific medical device usually can be acquired by modification efforts. That is, specialists from physicians to textile chemists and technologists can work as a team utilizing specific knowledge of their field to create an appropriate product. Shorter production cycles, higher flexibility and versatility, and lower production costs are some of the reasons for the popularity of nonwovens in medical textiles. Fig.52.4 is an artificial blood vessel made of industrial textiles.

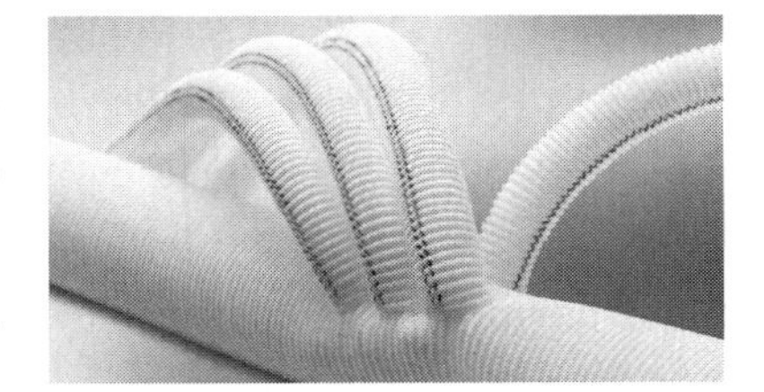

Fig.52.4 Artificial blood vessel

Military and defense textiles

Although the military strategies of the world powers are changing direction as a result of the end of the cold war, there will be a need for military textiles as long as there are military personnel (Fig.52.5). The recent trend in war fighting is towards quick deployment of agile forces made of fewer troops with increasingly sophisticated gear and weapons. The military textiles should meet the requirements of these new trends.

Soldiers are exposed to environmental conditions more than civilians. The differences between estimated outdoor-exposure lifetimes for civilian office workers, outdoor staff and peacetime soldiers are 3%, 8% and 20%, respectively, which explains why civilian clothing is not adequate for military use. Due to extremely strict requirements, the design, development and acceptance of a new military uniform takes a lot longer than it does in civilian clothing. Textiles in military and defense can be grouped into two categories: protective clothing and individual equipment, and textiles used in defense systems and weapons.

Fig.52.5 Helmet

Paper machine clothing

Large quantities of high value-added, high performance textile fabrics and felts are used in the paper industry. Although the average cost of paper machine clothing is only two cents for a dollar's worth of paper sold, paper machine clothing is one of the most critical factors in papermaking. In theory, paper can be made without a paper machine (which may cost several hundred million dollars) or without many of the chemicals involved, but it cannot be made without fabrics that are used for forming, pressing and drying of paper. Therefore, although the volume of paper machine clothing may be small, it is a stable and secure market due to the obvious need as long as the papermaking process is not revolutionized.

Safety and protective textiles

Safety and protective textiles refer to garments and other fabric-related items designed to protect the wearer from harsh environmental effects that may result in injury or death (Fig.52.6). It may also be necessary to protect the environment from people as in the case of clean rooms. Safety and protective materials must often withstand the effects of multiple harsh environments, such as ①extreme heat and fire, ②extreme cold, ③harmful chemicals and gases, ④bacterial/viral environment, ⑤contamination, ⑥mechanical hazards, ⑦ electrical hazards, ⑧radiation, ⑨vacuum and pressure fluctuations.

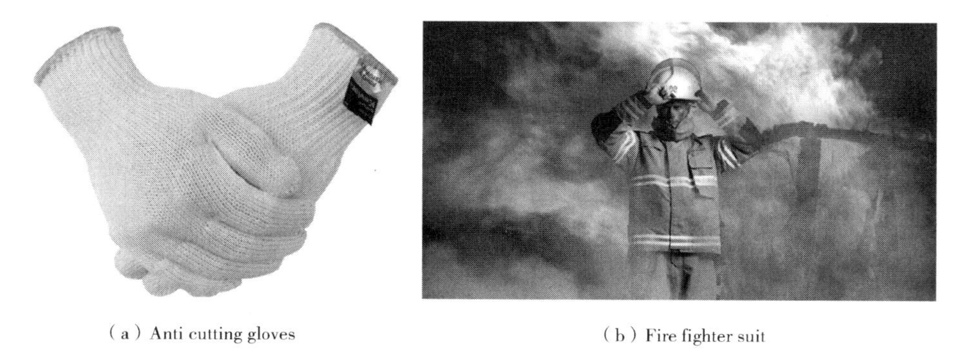

(a) Anti cutting gloves (b) Fire fighter suit

Fig.52.6 Anti cutting gloves and fire fighter suit.

Sports and recreation textiles

Manufacturers rely on sturdy technical textiles to create a broad variety of equipment used in indoor and outdoor sports and other recreational activities. High performance textile fibers and fabrics are used in uniforms, equipments and sport facilities (Fig.52.7). Textile manufacturers today are giving themselves an edge in the increasingly competitive sports and recreation market by developing their own signature products. Creating new fabrics is usually a lengthy and expensive process. Companies put their fibers and fabrics through rigorous tests to measure strength, abrasion resistance and breathability among other things. Several companies even have environmental labs where they can alter temperature and humidity conditions while athletes work out wearing the prototype garments.

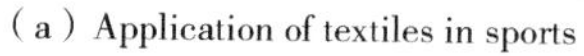
(a) Application of textiles in sports

(b) Recreation

Fig.52.7 Application of textiles in sports and recreation

Transportation textiles

Vehicles ranging from automobiles to railcars to boats and ships all rely on textiles in a variety of ways. Approximately 50 square yards of textile materials are used in an average car for interior trim (seating areas, headliners, side panels, carpets and trunk), reinforcement, lining, underlay fabrics, tires (Fig.52.8), filters, belts, hoses, airbags, sound dampening and insulation. Because of the overarching need for heat and fire resistance, whether inside the engine, passenger, or baggage areas, increasing use of fibrous and textile materials having higher levels of both mechanical performance, thermal insulation, and flame retardancy will continue to be a part of the necessary strategies adopted to increase passenger safety while maintaining or even reducing transport costs. It is important that in future vehicle design, especially if occupancy levels are increasing (as will fuel costs in the longer term), the role of technical textiles in ensuring both economic viability coupled with increased fire safety requirements will be increasingly recognized.

Fig.52.8 Tires reinforced by textiles

Textiles in agriculture

The main applications areas of textiles in agriculture include farming, animal husbandry and horticulture (Fig. 52. 9). The volume of special textiles that are manufactured for agricultural applications only is small compared to other areas of technical textiles. This does not mean that the use of textiles in agriculture is not significant. On the contrary, a wide variety of textile products that designed for general industrial applications are used in agriculture in great quantities. These products include hoses, conveyor belts, tires, composites, filters, textiles for hydraulic applications, etc. With the apparent exception of protective clothing for insecticides, farmers usually make use of existing fabrics to fit their needs.

Fig.52.9 Agriculture textiles

In addition to the items illustrated above, there are several other applications such as textiles in electronics, banners and flags, textile products, transport bags and sheets, fabrics to control oil spills, canvas and tarpaulins, ropes and nets, home and office furnishing and miscellaneous applications.

New Words and Expressions

preform 预制件
multiaxial 多轴向
thermoplastic 热塑性
thermoset 热固性
consolidation 固化
delamination 分层
tortuous 曲折的
permeable 透水的
plugging 堵塞
hydrogeology 水文地质
cytotoxicity 细胞毒性
prototype 原型，样板
interior trim 内饰
horticulture 园艺学
thermal insulation 隔热
flame retardancy 阻燃
miscellaneous 杂乱的

Notes to the Text

1. Textile fabrics are in fact a 3D network of fibers enclosing small pockets of void volume: 织物实际上是具有很多细小孔洞结构的三维纤维网络。
2. military personnel: 军事人员。
3. giving themselves an edge: 具有优势。

Questions to the Text

1. Describe the detailed classifications of industrial textiles.
2. For the industrial textiles in medical applications, how can we classify them?
3. What kind of special performances should the industrial textiles possess for the application in protective textiles?

语言微课堂

Lesson Fifty-Three

Differences Between Industrial Textiles and Non-industrial Textiles *

It is reasonable to state that, compared to textiles for apparel, there is a communication gap between industrial textile manufacturers and the end users, who are often in other industries. A typical example for this is the situation in geotextiles. The major geotextiles test methods and standards have been developed by civil engineers, most of the time without the involvement of textile engineers. This situation is generally true for almost all of the industrial textile products. Therefore, a direct feedback from the end user is not always readily available to the industrial textile manufacturer. The major reason for this might be the competitive nature of the markets and resulting confidentiality in other industries. For proper development of industrial textiles for a specific end use, the textile engineer should be knowledgeable enough about the application of his/her product. On the other hand, a civil or mechanical engineer should know something about design and manufacturing of industrial textiles so that he/she could give better insight to what needs to be developed.

思政微课堂：产业用纺织品的胜任潜质

There are several differences between industrial and traditional textiles that make industrial textiles unique:

(1) Application areas and end users of industrial textiles. Industrial textiles are usually used in non-textile industries. Traditional textiles are used mainly for clothing and home furnishing. A traditional textile product such as a garment is purchased and used by the consumer. Mostly the purchaser of an industrial fabric product does not use it for himself directly. Therefore, the direct user of industrial textiles is usually not the individual consumer. There is hardly any industry or human activity that does not involve the use of industrial textiles in one way or another. Almost every modern non-textile industry uses industrial textiles.

(2) Performance requirements. Depending on the application areas, industrial textiles are designed to perform for heavy duty and demanding applications. Failure of an apparel textile during use may cause some embarrassment for the user at worst. However, the consequences of failure of an industrial textile can be devastating. For example, failure of an air bag in a car accident or an astronaut's suit during a spacewalk may be fatal.

(3) Constituent materials. Due to different application areas and performance requirements, generally higher performance fibers, yarns and chemicals are used in industrial textiles. Materials with exceptional strength and resistance to various outside effects in turn give industrial textiles the

necessary strength and performance characteristics. In traditional clothing textiles, the physical performance requirements are less demanding compared to industrial textiles. However, comfort and appearance play a critical role in material selection for apparel. Functionality is the most important criteria for technical textiles; aesthetics, beauty, color, etc. are of secondary importance. For traditional textiles such as clothing, furnishing and household items, aesthetics and color are important properties along with functionality.

(4) Difference in manufacturing methods and equipment. Since industrial textiles consist of stronger constituent materials, manipulation of those materials is usually more difficult than weaker materials. Moreover, due to performance requirements, some industrial textiles may need to be built with high mass, therefore they are usually heavier than traditional clothing textiles. As a result, the manufacturing equipment and methods of consumer textiles may not be adequate for the manufacture of industrial textiles. For example, weaving of monofilament forming fabric, which is used on a paper machine to make paper, on a regular textile loom is impossible due to the very high weaving forces involved, resulting from the high strength and mass of polyester monofilament warp and filling yarns. The width of the forming fabric, which is determined by the width of the paper machine, can be up to 500 inches. For these reasons, specially built, very heavy and wide looms (up to 30 yards) are used in paper machine clothing manufacturing.

(5) Testing. Testing of industrial fabrics also presents another challenge. Quite often, once placed in an operation, industrial textiles cannot be changed or replaced easily. For example, geotextiles that are used under roads for reinforcement and stabilization cannot be replaced without complete destruction of the road. On the other hand, one cannot use a geotextile under a "test road" first because it may take many years to get the results. Failure of a geotextile in construction of a dam can cause devastating results.

Many times it is impossible to simulate the actual field conditions of industrial textiles in the laboratory. Therefore, empirical methods are either difficult to achieve or not as reliable for industrial textiles. As a result of these, the design engineer of an application involving industrial textiles quite often has to rely on laboratory test results. This increases the need for accuracy and reliability of the test results for industrial textiles. Conventional textile test methods are often not suitable for industrial textiles. Therefore, new test methods and procedures have been developed for industrial textiles. Simulation and modeling of field conditions with computer aided design systems is also becoming a more common practice to be able to determine the optimum structure and performance of industrial textiles for a particular end use. In addition to physical properties, the "performance" or "quality" of clothing textiles depends on other variables such as consumer perception and taste. Therefore, evaluation and specification of clothing textiles is rather subjective and may be difficult.

For most of the tests of industrial textiles, the performance and evaluation are well defined based on the end use. This fact makes it relatively easy to develop exact measuring and testing methods for industrial textiles. As a result, development of performance data bases is also easier.

(6) Life expectancy. In general, industrial textiles are expected to last much longer than traditional consumer textiles. Unlike consumer textiles, fashion trends do not play a role in the life of traditional textiles. In large structures such as buildings, highways, stadiums or airplanes, textiles are expected to last many years. Although longer life is desirable for industrial textiles, it is not always attainable and some industrial textiles may have a shorter useful life than traditional textiles. Moreover, in some applications, it is desirable to control or limit the life of an industrial textile. For example, some textile materials that are used inside or outside the human body during surgery are expected to degrade after they complete their task (e. g., when the natural organ or tissue grows and gains enough strength) which may take a few weeks or months. The dissolved material is carried away by the body fluids and discharged from the body.

(7) Price. As a result of all the wonders that industrial textiles offer, it is expected that they come at a higher initial cost than traditional textiles. However, considering the long life and benefits they offer and contributions they make to the national infrastructure and economy, the cost of industrial textiles should not be a concern in the overall picture. In fact, when properly constructed and applied, industrial textiles can be used in place of more expensive materials and therefore may bring substantial savings to the overall economy in the long range.

New Words and Expressions

aesthetics 美学
dam 坝
subjective 主观的，个人认为的
expectancy 期望值
infrastructure 基础设施

Notes to the Text

1. a communication gap：指一个可以沟通的鸿沟。
2. Therefore, empirical methods are either difficult to achieve or not as reliable for industrial textiles：因此，对于产业用纺织品测试的传统方法要么就难以实现，要么就并不可靠。

Questions to the Text

Decribe the detailed differences between industrial textiles and traditional textiles.

Lesson Fifty-Four

Textile Reinforced Composite Materials*

Textile reinforced composite materials are part of the general class of engineering materials called composite materials. It is usual to divide all engineering materials into four classes: metal, polymers, ceramics and composites. Composite materials are characterized by being multiphase materials within which the phase distribution and geometry has been deliberately tailored to optimize one or more properties. This is clearly an appropriate definition for textile reinforced composites for which there is one phase, called the matrix, reinforced by a fibrous reinforcement in the form of textile.

In principle, there are as many combinations of fiber and matrix available for textile reinforced composites as there are available for the general class of composite materials. In addition to a wide choice of materials, there is the added factor of the manufacturing route to consider, because a valued feature of composite materials is the ability to manufacture the article at the same time as the material itself is being processed. This feature contrasts with the other classes of engineering materials, where it is usual for the material to be produced first (e. g. steel sheet) followed by the forming of the desired shape.

The full range of possibilities for composite materials is very large. In terms of reinforcement we must include glass fibers, carbon fibers, ceramic fibers and aramid fibers. Matrices include wide ranges of polymers (polyesters, nylons, etc.), metals (aluminum alloys, etc.) and ceramics.

The market for composite materials can be loosely divided into two categories: reinforced plastics based on short fiber E-glass reinforced unsaturated polyester resins (which account for over 95% of the volume) and advanced composites which make use of the advanced fibers (carbon, aramid, etc.), or advanced matrices (e. g. high-temperature resistant polymer, metallic or ceramic matrices).

Textile reinforced composites have been in service in engineering applications for many years in low cost applications. While there has been a continual interest in textile reinforcement since around 1970, and increasingly in the 1980s, the recent desire to expand the composite usage has had a dramatic effect on global research into, and usage of, textile reinforcement. In addition to the possibility of a range of applications for which textile reinforcement could replace current metal technology, textile reinforcement is also in competition with relatively mature composite technologies which use more traditional methods of production. This is because textile reinforced composite materials show potential for reduced manufacturing costs and enhanced processability, with more than adequate, or in some cases improved, mechanical properties. Those economic entities within which composite materials have been well developed, notably the European Economic Community (with about 30% of the global composite usage), the USA (with about 30%) and Japan (with about 10%) have been a growing interest in textile reinforcement in the 1990s, with China, Russia, South Korea, India, Israel and Australia being additional major contributors. In the last years of the 20th century, conferences devoted to composite materials always had important sections on textile reinforcement.

Of the available textile reinforcements (woven, braided, knitted, stitched), woven fabric reinforcement for polymer matrices can now be considered to be a mature application, but many textiles are still the subject of demonstrator projects. Several textile techniques are likely to be combined for some applications. For example, a combination of braiding and knitting can be used to produce an I-shaped structure.

For structural applications, the properties which are usually considered first are stiffness, strength and resistance to damage and crack growth.

The 1990s saw a growing mood of cautious optimism within the composites community worldwide that textile based composites will give rise to new composite material applications in a wide range of areas. Consequently, a wide range of textile reinforced composites are under development/investigation or in production. Textile reinforcement is thus likely to provide major new areas of opportunity for composite materials in the future.

New Words and Expressions

ceramic　陶瓷的
tailor　制作
matrix　基体(复数为 matrices)
article　物品，产品
processability　可加工能力
entities　(entity 的复数形式)实体
European Economic Community　欧洲经济共同体
stitched　缝合的

Notes to the Text

1. Composite materials are characterized by being multiphase materials within which the phase distribution and geometry has been deliberately tailored to optimize one or more properties: 多相性是复合材料的特点，相的分布和几何特性可以人为地控制使一种或多种性质达到最佳。
2. There are as many combinations of fiber and matrix available for textile reinforced composites as there are available for the general class of composite materials: 纤维与基体的结合方式很多，就像目前能得到的复合材料的种类那样多。
3. A valued feature of composite materials is the ability to manufacture the article at the same

time as the material itself is being processed：复合材料一个有价值的特点是生产材料的同时产品也生产出来了。既在生产材料的同时，产品也同时完成了。
4. The full range of possibilities for composite materials is very large：用来制作复合材料的原料范围较广。
5. unsaturated polyester resin：不饱和聚酯树脂。
6. which account for over 95% of the volume：（玻璃纤维增强材料）占纺织复合材料的 95%。
7. with more than adequate：较好的。
8. but many textiles are still the subject of demonstrator projects：但是许多纺织材料依然用于展示。意思是纺织材料的应用还不广泛。
9. a growing mood of cautious optimism：一种逐渐增长而且谨慎的乐观情绪。

Questions to the Text

1. What are the major engineering materials?
2. Describe something about the usage of the composite materials in different countries.

Lesson Fifty-Five

Textiles in Transportation*

语言微课堂

专业微课堂：汽车用碳纤维复合材料

Transportation is the largest user of technical textiles. Textiles provide a means of decoration and a warm soft touch to surfaces that are necessary features for comfort, but textiles are also essential components of the more functional parts of road vehicles, trains, aircraft and sea vessels. Textiles in transportation are classed as technical because of the very high performance specifications and special properties required. Seat covering, for example, are not easily removable for cleaning and indeed in automobiles they are fixed in place and must last the lifetime of the car without ever being put in a washing machine. In trains, aircraft and passenger vessels they are exposed to much more rigorous use than domestic furniture. In addition they have to withstand much higher exposure to daylight and damaging ultraviolet radiation and because they are for public use they must satisfy requirements such as flame retardancy.

In more functional applications, textiles are used as tires, heater hoses, battery separators, brake and clutch linings, air filters and drive belts. Fiber reinforced composites are replacing metallic components with considerable benefits, especially saving in weight.

The most familiar technical textile in transportation is car seat fabric which is amongst the largest in volume and is growing annually in the developing world. Car seat fabric requires considerable technical input to produce both the aesthetic and also the very demanding durability requirements. The processes developed for car seat fabric and the technical specifications provide some indication of the requirements for seat materials in other transport applications.

In all transportation applications certain important factors recur, like comfort, safety and weight saving. In public transportation situations as far as textiles are concerned safety means reduced flammability. Environmental factors have also become important and these have influenced the transportation textile industry in a number of ways including design, choice of materials and manufacturing methods.

Reduced flammability properties are understandable considering the restrictions on escape routes especially in the air and at sea. Flame retardancy requirements of private cars are not especially high but are necessary for passenger trains and standards are increasing for passenger coaches. Transportation disasters which frequently become headline news are impetus for increased flame retardancy standards and improvements in public safety.

Resistance to sunlight is perhaps the most important property a fabric must have. Choice of wrong fabric can lead to breakdown of the seat cover within weeks, depending on the intensity of the sunlight, the UV radiation. Actual degradation by UV radiation is influenced by the thickness of the yarn, the thicker the better because less radiation will penetrate into the center. This is particularly the case for nylon yarns. Significant improvements in UV radiation resistance can be obtained by addition of certain chemicals that are UV absorbers and these are used extensively with polyester, nylon and polypropylene for transportation applications. UV absorbers in nylon are usually added to delustered yarns.

Seating fabric needs to be of the highest standard of abrasion resistance. Only polyester, nylon and polypropylene are generally acceptable, although wool is used in some more expensive vehicles because of its aesthetics and comfort. Wool has other properties such as non-melting and reduced flammability which make it suitable for aircraft seats. Fabric abrasion is influenced by yarn thickness, texture, cross-section and whether spun or continuous filament.

The whole area of transportation is growing with increasing trade between all the nations of the world generating higher volume both in freight and also commercial passenger travel. Leisure travel is also increasing dramatically with larger incomes, increasing leisure time and increasing interest in foreign countries; the largest growths are expected in air travel.

The stresses of modern living require transportation interiors to be more pleasing and relaxing, to ease traveling and to make journeys more enjoyable. Indeed the various forms of transport now compete with each other for passengers. For many national internal journeys the traveling times and costs of, say air and rail are very similar from city to city.

A further requirement of textiles involving passenger transport is cleanability. The only opportunity for servicing is at the end of a journey and just before the start of the next one. Cleaning must be done as quickly as possible. Easy care and maintenance are very important; dirty carpets and seats would annoy passengers.

Technical textiles are relatively newcomers to the textile industry, which is probably one of the world's oldest industries, but there are still opportunities to learn from more traditional methods with other industries.

Although textiles have been used in some car seats since the innovation of the car, widespread use has only occurred since the mid-1970s. The technology and manufacturing methods are still on learning period compared to other sectors of the textile industry. Fabric car seats could still benefit from certain developments and processes that have been available to the garment and finishing industry for many years, for example advanced finishing techniques providing softer handle and touch, antistatic finishes, specialist yarns and techniques for improved thermal comfort.

New Words and Expressions

technical textile 技术纺织品，工业用纺织品，产业用纺织品
functional part 功能零件
sea vessel 船
passenger vessel 客船
flame retardancy 阻燃性

battery separator 电池内隔板
clutch lining 制动器内衬
passenger coach 长途公共汽车
deluster 退光，消光
freight 货物，货运
interior 内部，内部装饰
cleanability 可清洁性
specialist yarn 特种纱线

Notes to the Text

1. Car seat fabric requires considerable technical input to produce both the aesthetic and also the very demanding durability requirements：为了美观和较好的耐久性，在汽车坐垫织物的生产中需要较大的技术投入。
2. the restrictions on escape routes：指受到限制的疏散通路。
3. deluster 为美国用法，英国用法为“delustre”。
4. non-melting：不熔化。
5. commercial passenger travel：指商业目的的旅行。
6. stresses of modern living：当代生活带来的压力。
7. For many national internal journeys the traveling times and costs of, say air and rail are very similar from city to city：许多国内城市之间的旅行，乘飞机和火车所用的时间和费用是相近的(指竞争带来的结果)。
8. The only opportunity for servicing is at the end of a journey and just before the start of the next one：“Servicing”是指对运输工具内部的整理和清扫。
9. easy care：容易清洗和保养。
10. Technical textiles are relatively newcomers to the textile industry, which is probably one of the world's oldest industries：句中“which”是指“the textile industry”。
11. Fabric car seats could still benefit from certain developments and processes that have been available to the garment and finishing industry for many years：指汽车坐垫织物的生产可以从有多年经验的服装和整理方面学到很多东西。

Questions to the Text

1. What are the main requirements for the car seat textiles?
2. Why flame retardant is more important in the public transportation?
3. What materials may be selected for the public seat cover? Why?

语言微课堂

Lesson Fifty-Six

Geotextiles *

One of the most rapidly growing uses for textiles is in geotextile applications. Geotextiles include textiles for ground stabilization, erosion control, drainage, asphalt and moisture proofing. Both woven and nonwoven fabrics are used in geotextile applications. These fabrics are made of such fibers as polypropylene, olefin, polyester, nylon, and acrylic. Nonwoven fabrics exceed woven in use. The most important fiber, or most used, is polypropylene. Polyester would provide the best performance, but it is considered too costly for widespread use; acrylics are also considered too expensive despite the fact that they have superior properties for such uses. Probably the most common use of geotextiles is in the area of soil stabilization, erosion control, and drainage in ground installations. They serve as a separation layer to form a stable boundary between layers of soil to maintain the integrity, character and performance properties of each layer. This usage occurs in such situations as constructing an embankment of soil or laying ballast for a road or railroad. These textile materials act as a barrier to the movement of soil particles but permit the flow of water necessary for adequate drainage.

Drainage can occur from one soil layer to another through a geotextile layer, or occur within a single plane or layer where the geotextile serves as the drain vehicle. The geotextile material reinforces a structure and reduces erosion by equalizing stresses over a wide area.

Textiles are used as a support or stabilizer for concrete in the construction of dams, canals, drainage systems and retaining walls for ponds or lakes. They are used in road construction in various ways. They may be laid over the soil and rock base to help segregate the subsoil and the fill, and they may be laid over the base directly under the surface asphalt layers. The former use provides for equalization of stress as well as segregation of subsoil and fill; the later reduces cracking of the road base and asphalt surface and reduces the amount of asphalt required for good stability of the road surface. Both uses tend to improve drainage.

The three main properties which are required and specified for a geotextile are its mechanical responses, filtration ability and chemical resistance. These are the properties that produce the required working effect. They are all developed from the combination of the physical form of the polymer fibers, their textile construction and the polymer chemical characteristics. For example, the mechanical response of a geotextile will depend upon the orientation and regularity of the fibers as well as the type of polymer from which it is made. Also, the chemical resistance of a geotextile will depend upon the size of the individual component fibers in the fabric, as well as their chemical composition—fine fibers with a large specific surface area are subject to more rapid chemical attack than coarse fibers of the same polymer.

Mechanical responses include the ability of a textile to perform work in a stress environment and its ability to resist damage in a serious environment. The ability to perform work is fundamentally governed by the stiffness of the textile in tension and its ability to resist creep failure under any given load condition.

The weight or area density of the fabric is an indicator of mechanical performance only within specific groups of textiles, but not between one type of construction and another. For example, within the overall range of needle punched continuous filament polyester fabrics, weight will correlate with tensile stiffness. However, a woven fabric with a given area density will almost certainly be much stiffer than an equivalent weight needle punched structure. Clearly the construction controls the performance. Therefore, it is impossible to use weight alone as a criterion in specifying textiles for civil engineering use. However, in combination with other specified factors, weight is a useful indication of the kind of product required for a particular purpose.

The breaking strength of a standard width of fabrics is universally quoted in the manufacturers' literature to describe the strength of their textiles. This is of very limited use in terms of design. No designer actually uses the failure strength to develop a design. Rather, a strength at a given small strain level will be the design requirement.

Creep can cause the physical failure of a geotextile if it held under too high a mechanical stress. It has been found that in practical terms, both polyester and polyethylene will stabilize against creep if stress levels can be maintained at a sufficiently low level. Although polypropylene does not seem to stabilize at any stress level, its creep rate is low at small stresses that a "no creep" condition may be considered to exist in practice.

The filtration performance of a geotextile is governed by several factors. To understand this, it is essential to be aware that the function of the textile is not truly as a filter in the literal sense. In general, filters remove particles suspended in a fluid, for example, dust filters in air-conditioning units, or water filters, which are intended to remove impurities from suspension. Quite the opposite state exists with geotextile filters. The geotextile's function is to hold intact a freshly prepared soil surface, so that water may pass from the soil surface and through the textile without breaking down the surface.

Filtration is one of the most important functions of textile used in civil engineering earthworks. It is without doubt the largest application of textiles includes their use in the lining of ditches, beneath roads, for building basement drainage and in many other ways.

Of all the varied uses for geotextiles, only in a reinforced soil mass is there no beneficial filtration effect. In just about all other applications including drains, roads basements and embankment supports, the geotextile will play a primary or secondary filtering function.

The permeability of geotextiles can vary considerably, depending upon the construction of the fabric. Various national and international standards have been set up for the measurement of permeability that is required, most often at right angles to the plane of the textile, but also along the plane of the textile. It is important in civil engineering earthworks that water should flow freely through the geotextile, thus preventing the build-up of unnecessary water pressure. Fig.56.1 shows the usage of the geotextiles in Beijing Daxing International Airport.

Geotextiles are rarely called upon to resist extremely aggressive chemical environment. Particular examples of where they are, however, include their use in the basic layers of chemical containers or waste disposal sites. This can happen if leaks occur. Another example might be the use of textiles in contact with highly acidic soils, where in tropical countries, pH down to 2 have been encountered. In industrialized countries where highways are being built through highly polluted areas, geotextile can come into contact with serious environments.

专业微课堂：
土工布

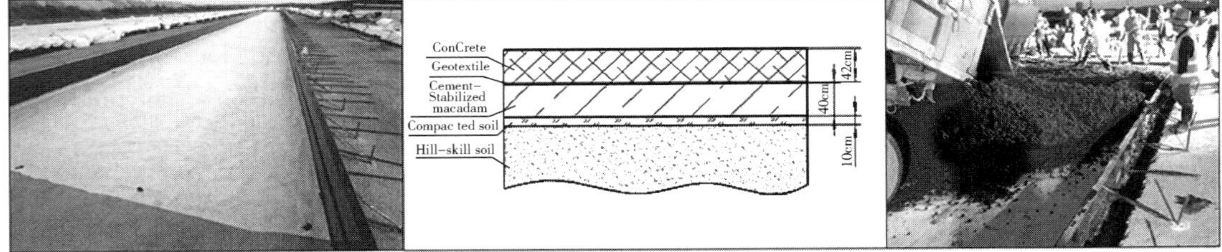

Fig.56.1 Usage of geotextiles (Source: http://www. tdf. com. cn/product/33. html)

Although the chemical mechanisms involved in fiber degradation are complex, there are four main agents of deterioration: organic, inorganic, light exposure and time change within the textile fibers.

Organic agents include attack by micro-organisms. This is not considered to be a major source of deterioration. Micro-organisms may damage the textiles by living on or within the fibers and producing detrimental byproducts.

Inorganic attack is generally restricted to extreme pH environments. For example, polyester will be attacked by pH levels greater than 11, but these are rare.

Ultraviolet light will deteriorate geotextile fibers if exposed for significant periods of time, but laboratory testing has shown that fibers will deteriorate on their own with time, even if stored under dry dark cool conditions in a laboratory. Therefore, time itself is a damaging agent as a consequence of the fiber deterioration.

New Words and Expressions

erosion 腐蚀，侵蚀
fill 填充土
concrete 混凝土
dam 水坝
failure 破坏
correlate 相关
literal 实际的
suspend 悬浮
earthwork 土木工程
ditch 沟渠
tropical 热带的
chemical mechanism 化学机理
detrimental 有害的，不利的
byproduct 副产品

Notes to the Text

1. erosion control：控制水土流失。
2. Nonwoven fabrics exceed woven in use：非织造布的应用超过了机织物。在“woven”后面省略了“fabrics”。
3. road or railroad：可译为“公路或铁路”。
4. but permit the flow of water necessary for adequate drainage：“necessary for adequate drainage”做“flow of water”的后置定语。
5. The three main properties which are required and specified for a geotextile are：句中“are required and specified”指土工织物特别需要的。
6. Also, the chemical resistance of a geotextile will depend upon the size of the individual component fibers in the fabric：句中“size”指纤维的粗细。
7. specific surface area：比表面积。
8. stress environment：具有应力的环境。
9. The weight or area density of the fabric is an indicator of mechanical performance only within specific groups of textiles：句中“only within specific groups of textiles”指特指某一种纺织品。
10. but not between one type of construction and another：但不是两种结构之间(的比较)。
11. manufacturers' literature：生产商提供的说明书及相关文件。
12. This is of very limited use in terms of design：这些在产品设计中的用途是很有限的。
13. No designer actually uses the failure strength to develop a design. Rather, a strength at a given small strain level will be the design requirement：实际上设计人员不使用破坏强度进行产品设计，而是使用产生较小应变时的强度进行设计。
14. it is essential to be aware that the function of the textile is not truly as a filter in the literal sense：句中“in the literal sense”指“在实际上”。
15. Of all the varied uses for geotextiles, only in a reinforced soil mass is there no beneficial filtration effect：对于所有土工织物来说，仅仅用来加固土层的织物没有过滤的效果。

Questions to the Text

1. What are the main usages of geotextiles?
2. Cite a few examples you know of the end use of geotextiles.
3. What are the main requirements for the geotextiles?
4. What factors influence the mechanical performance of the geotextiles?
5. Why filtration of the geotextile is important in some cases?
6. What will cause the geotextile to deteriorate?

Lesson Fifty-Seven

Future Textiles

语言微课堂

In the hierarchy of human needs, clothing is almost as important as food. We have come a long way in the last 40,000 years and witnessed the continual progression of textiles and clothing from those first, basic needs.

The textile industry has already experienced two major phases of transformation that have had a large impact on textiles: the Industrial Revolution in the period 1750—1850 and the development of synthetic/high performance fibers in the 20th century.

It is clear that textile and apparel developments have enabled individuals to reach increasingly higher goals of human performance; prevented injury where once a soul would have undoubtedly perished; helped with surveillance, communication, and navigation techniques; kept us physically safe; and ensured our well-being. For example, we now have protective vests for the police and the military. Army helmets and vests are mostly made from Kevlar, a derivate of polyamide (nylon), with strength-to-weight ratio stronger than that of steel. The advantage of this fiber is its lightweight properties with its high strength combined to produce a garment that can stop a bullet from certain ranges while also being light enough for the military to wear. Fabrics such as Nomex are extensively used for pilots in military aircraft, firefighters, and formula one racing. Garments for racing were once made from 100% wool, and this material was used because one of wool's unique properties is that it does not catch fire but chars giving some protection to the driver, if a fire occurs in the driver's cockpit. Similarly, it also offers protection to the pilots in case of being shot down, reducing the risk of injury in a fire or even death. These fibers are known as para-aramids and aramids respectfully. As the range of fabric types and demands from the seams and textile properties have evolved, the range of methods for making innovative fabrics and joining of the seams has also increased.

Textiles are engineered to fulfill a purpose; the next generation of high-performance textiles and apparels will provide more and complex functionalities for technical applications.

New changes can be expected in the near future with the introduction of the next generation materials such as smart textiles and nanomaterials, bringing new functionalities and further increasing the level of performance. For instance, smart materials and wearable electronics allow textiles to become active and reactive. Clothes are in contact with an extensive part of the body and we wear it every day. It is flexible, low-cost, versatile, light, and an ideal candidate for wearable electronics. The evolution of electronics has revolutionized the textile industry. The previously lifeless textile product can provide active cooling and/or heating, adapt its breathability to environmental conditions, guide visually impaired people, allow integrated position location and communication, offer physiological monitoring, release active compounds and drugs at a controlled rate, monitor the structural health of composite structures, and harvest and transfer energy for example.

Recently, textiles with the use of electronic components have been attracting an intense research activity and commercial companies, mainly driven by the advances of the electronic industry in terms of flexible electronics, thin film technology, 3D printed electronics, miniaturization, low-power consumption, wireless charging, skin sensor technology, encapsulation, mobile apps, and open innovation. Many international Internet or technology giants, such as Google, Apple and Huawei, published a few patents related to textile electronics. Google and Levi's launched together in 2014 the Project Jacquard. This project aims to make Jacquard garments interactive, allowing designers and developers to build connected touch-sensitive textiles into their own products. Google and Levi's are planning to launch soon the Google Trucker Jacket from this partnership. Google has also recently partnered with a company, to take its Project Jacquard for the healthcare industry uniforms.

The future trends are towards low-power consumption and flexible printable electronics, such as the stretchable electronic ink and flexible electronics. Another trend to be embraced by textiles is 3D printed electronics.

专业微课堂：
鲨鱼皮泳衣

Nanotechnology is another science that is finding its way into the textile industry, either through the use of nanofibers, the surface modification or coating of fibers and textile structures using nanomaterials, or the addition of nanofillers for the production of nanocomposite fibers and filaments. Nanotechnology offers great opportunities to improve the properties and performance of existing textile products and develop completely new functionalities.

A second driver of change in the textile and apparel industry is the shift towards more environmentally friendly manufacturing. This trend is fueled by new regulations aimed at better controlling the amount of chemicals used in textile processes and limiting those that are the most toxic.

思政微课堂：
纺织智能工厂

Additionally, with the rapid development of the industry and the rapid increase of consumption, the disposal of waste textiles also comes along. Waste chemical fiber textiles are treated as waste for simple disposal such as landfill or incineration, which not only seriously pollutes the environment, but also causes great waste of resources. Seeking for recycling has become an urgent need for the development of world's textile industry.

New Words and Expressions

progression 进展，发展

surveillance 监督，监视

vests 背心

helmets 头盔

polyamide 聚酰胺

cockpit 驾驶舱

para-aramids　对位芳纶
wearable electronics　可穿戴电子产品
nanofillers　纳米填加物
landfill　垃圾填埋
incineration　焚化，垃圾焚烧
recycling　循环利用

Notes to the Text

1. In the hierarchy of human needs, clothing is almost as important as food：在人类基本需求的层次中，服装与食物几乎同等重要。
2. strength-to-weight ratio：直译为重量强度比，也称比强度（specific strength）。
3. formula one racing：一级方程式赛车。
4. Nomex：诺梅克斯，一种间位芳纶，学名为聚间苯二甲酰间苯二胺（国内称芳纶1313），纤维具有优异的耐高温性和阻燃性，是杜邦公司生产纤维的商品名。
5. wireless charging：无线充电。
6. surface modification：表面改性。
7. nanocomposite fibers and filaments：纳米复合纤维和长丝。

Questions to the Text

1. Discuss the possible future textiles and apparels with your friends.
2. Write a mini-review paper on recycling of textiles and apparels.

Vocabulary

5-shaft satin　五枚缎纹(32)
a guide needle　引导针(15)
a hollow spindle　空心锭子(15)
abrasion resistance　耐磨强度(32)
absorption　吸收(46)
accessories　服装的附件(44)
acetate fiber　醋酯纤维(8)
acetate　醋酯(17)
acetic anhydride　乙酸酐(9)
acid sweat resistance　耐酸汗渍(47)
acrylic　腈纶(9)
additive tensioner　倍加式张力器(25)
additive　助剂(27)
adept　能手，内行(39)
adhesive　黏合剂(27)
adsorption　吸附，吸收(46)
aerodynamically　空气动力学地(39)
aerosols　气溶胶(40)
aerospace　航空航天(51)
aesthetic　美的，美学的，审美的(47)
aesthetics　美学(53)
affinity　亲和性(38)
aggregate　混凝土粒料(41)
aging　老化(6)
air bonding　热风黏合(39)
air jet loom　喷气织机(31)
air permeability　透气性(19)
airborne　空气传播的(40)
air-laid　气流成网(39)
airport runway　飞机跑道(41)
A-line　A 字形裙(43)
alkaline sweat resistance　耐碱汗渍(47)
alloy　合金(50)
aluminum chloride　氯化铝(5)
amendment　修改(48)
amine oxide　氧化胺(9)
ammonium peroxysulfate　过硫酸铵(50)
amorphous area　非结晶区，无定形区(9)
amorphous　无定形的(8)
angora goat　安哥拉山羊(48)
angular velocity　角速度(25)
anisotropic　各向异性的(39)
antenna　(antennae)天线(49)
anthropologist　人类学家(1)
antibacterial　抗菌的(11)
antimicrobial　抗菌的(11)
antiperspirant　除汗剂(5)
antique satin　仿古缎(32)
antistatic device　抗静电装置(26)
apparel　服装；衣服(1)
applique　嵌花(43)
apprenticeship　学徒年限，学徒身份(23)
apron　胶圈，皮板输送带(14)
aramid fiber　芳纶(49)
aramid　芳香族聚酰胺(39)
articial fibers　人造纤维(1)
artificial leather　人造皮革(10)
artificial kidney　人工肾(51)
asbestos　石棉(1)
Asia Minor　小亚细亚(1)
asphalt　沥青(41)
assemblage　集合体(14)

atlas lapping　经缎(36)
attenuate　使变细(14)
autoclave　高压灭菌器；高压锅(50)
auxiliary cam shaft　辅助踏盘轴，辅助凸轮轴(30)
auxiliary jet　辅助喷嘴(31)
back rest　后梁(29)
back roll　后梁(29)
bacteria　(bacterium 的复数形式)细菌(6)
bagging　打包布，麻袋布(7)
balanced fabric　平衡织物，经纬向参数相同的织物(45)
balling　起球(28)
ballooning　气圈(25)
barrier　阻挡层(17)
basic weave　基础组织(30)
bast fiber　韧皮纤维(6)
batiste　细薄织物，法国上等细亚麻织物(32)
batt　棉絮；絮垫(39)
battery separator　电池内隔板(55)
beam creel　经轴架(27)
beam warping　轴经整经(26)
bearded needle　钩针(35)
beat up　打纬(27)
beating-up　打纬(29)
beige　米色(7)
bell bottom　裤子喇叭脚口(43)
bengaline　罗缎，孟加拉织品(32)
Bermuda shorts　百慕大短裤(43)
bicomponent　双组分(16)
bilobal　双叶形的(9)
bin　箱，池，仓(49)
biocide agent　杀菌剂(11)
blazer jacket　运动夹克，宽松外衣，便装(43)
bleed　渗色(38)
blender　混棉机，混合机(14)
blending　混纺(2)
block copolymer　嵌段聚合物(11)
blouse　女衬衫，罩衫(17)
boat rigging　船缆，船索(49)
bobbin　筒管(14)
body armor　护身盔甲(49)
body　身骨(16)
boll　植物的铃(6)
bombyx mori　家蚕(1)
bouffant　蓬松裙(43)
bracing belts　船用索带(49)
braid　编织(1)
break spinning　自由端纺纱，气流纺纱(15)
breaker-drawing　预并条机(14)
breaking extension　断裂伸长(18)
breaking length　断裂长度(18)
breaking strength　断裂强力(18)
breed　育种，品种(3)
bridal satin　婚服缎(32)
bulk　蓬松(17)
bulkiness　蓬松度(13)
bulky　蓬松的(17)
bulletproof vest　防弹背心(49)
burlap　粗麻布(7)
burning test　燃烧测试(13)
burr　草籽，草刺(23)
burst rod　分绞棒(27)
bushing　套筒，套管(49)
bustle　裙撑(43)
butt　针踵(35)
butterfly wings　蝴蝶翅膀(50)
byproduct　副产品(56)
cable　粗绳索(18)
calender bonding　热轧黏合(39)
cam shaft　织机中轴，织机踏盘轴，织机凸轮轴(30)

cam track　三角轨道(35)
can　条筒(14)
canvas　帆布(51)
capacitance　电容(19)
capstan tensioner　柱式张力器，倍积式张力器(25)
carbon fiber　碳纤维(49)
carboxymethyl cellulose(CMC)　羧甲基纤维素(27)
card clothing　针布(14)
card unit　梳理机构(14)
carded　粗梳的(14)
carder　梳理机(24)
carding oil　梳毛油(23)
carding　梳理，粗梳(6)
care label　使用说明标签(5)
care　保养，护理(42)
carpet backing　地毯背用布(7)
carton　纸板箱(14)
casein　乳酪(9)
cashmere goat　山羊(48)
caustics　腐蚀剂(41)
cellulose acetate　醋酯纤维素(9)
cellulose　纤维素(2)
cellulosic fiber　纤维素纤维(5)
cement　黏合(27)
centrifugal force　离心力(25)
ceramic　陶瓷，陶瓷制品(39)
chain return mechanism　链条式片梭返回装置(31)
chambray　钱布雷平布，色经白纬平布(32)
cheesecloth　干酪包布(32)
chemical fiber　化学纤维(1)
chemical mechanism　化学机理(56)
chiffon　雪纺绸，薄纱(32)
chiral　手性的(50)
chlorine　氯(5)
choice fiber　上等纤维(6)
choice product　精品(5)
chromatic fiber　变色纤维(12)
chromatograph　色谱仪(13)
civil and mechanical engineering　土木和机械工程(51)
civil engineering　土木工程(41)
cleanability　可清洁性(55)
cleaning　除杂(14)
clear shed　清晰梭口(30)
clearing　退圈(35)
closed lap　闭口线圈(34)
cloth roll　卷布辊(29)
cloth roller　卷布辊(29)
cloth　布，织物(1)
clothing wool　粗梳用毛(22)
clo　克罗值(服装保温能力单位)(46)
cluster　一簇羊毛(24)
clutch lining　制动器内衬(55)
cobweb　蜘蛛网(49)
cockpit　驾驶舱(57)
cocoon　蚕茧(1)
collar　领圈(44)
collection　接收(31)
color change　变色(47)
color fastness　染色牢度(47)
color intensity　颜色强度
color value　色值(38)
colorant　颜料，染料(2)
comb　梳理(6)
combed yarn　精梳纱(14)
comber　精梳机(14)
combined tensioner　联合式张力器(25)
combing wool　精梳用毛(22)
compactness　紧密度(45)
compatibility　适应性，相容性(27)
complex weave　复杂组织(30)
composite material　复合材料(31)

compound needle　复合针(35)
concrete　混凝土(56)
conduction　传导(46)
cone　圆锥形筒子(25)
confusor　管道片(喷气织机防气流扩散的构件)(31)
consecutive　连贯的(50)
consolidation　固结；加固；固化(39)
contaminant　沾染物(9)
contort　歪斜(31)
contraction　收缩(8)
convection　对流(46)
conventional loom　常规织机，有梭织机(31)
conveyor belt　传送带(14)
conveyor　输送机；传送带(39)
convolution　扭曲(2)
cordage　绳索(7)
core-spun yarn　包芯纱(16)
correlate　相关(56)
corrosive　腐蚀的(41)
cortex　角质，皮质(4)
cortical cell　角质细胞，皮质细胞(4)
cosmetic　化妆品(44)
cotton　棉(1)
counteract　抵消，中和，阻碍(17)
counterelectrodes　对电极(12)
course　横列(34)
covering power　覆盖系数(4)
coverspun　包绕纺纱(16)
crankshaft　曲拐轴(30)
crease　褶裥(39)
creeling　换筒(26)
creep　蠕变，塑性变形(28)
crepe yarns　绉纱(42)
crepe-back satin　绉缎(32)
crimp　卷曲(17)
crisp　挺爽(2)
cross-lapping　交叉铺网(39)
cross-linkage　交键(4)
crust　硬皮(8)
crystallinity　结晶度(7)
cuprammonium rayon　铜铵人造丝(9)
curl　卷曲(18)
cushioning　垫子(41)
custom made　定制的，定做的(44)
cuticle　表皮(2)
cuticle　角质层(4)
cylinder　锡林(14)
cytotoxicity　细胞毒性(52)
dam　坝(53)
Daolian Drawing　捣练图(1)
deep sea workstation　深海工作站(49)
degradation　降解(2)
degree of crystallinity　结晶度(9)
degree of polymerization　聚合度(9)
degum　脱胶(7)
delamination　分层(52)
delustered　消光的(9)
deluster　退光，消光(55)
delustrants　消光剂(13)
denier　旦，旦尼尔(5)
dent　筘齿、筘隙(29)
deodorant　除臭剂(5)
deplete　耗尽(7)
depletion　耗尽(25)
deposit　存留，囤积(3)
desorption　水分释放(4)
deteriorate　损害；恶化(3)
detrimental　有害的，不利的(19)
dew retting　露水沤麻(6)
dialysis　透析(51)
diaper　尿片，纸尿裤(10)
dielectric constant　介电常数(19)
differential dye technique　差异染色技术(15)

diffuse　扩散，漫射，使分散(17)
diffusion　扩散(46)
diisocyanate　二异氰酯(10)
dilute sulfuric acid　稀硫酸(6)
dipole　偶极(50)
direct take-up　直接卷取(29)
dirigible　飞船(51)
dirndl skirt　紧身连衫裙(43)
discard　丢弃(39)
discoloration　脱色(6)
disposable　用可弃的(41)
disposal　处理，排放(3)
disrobe　脱衣(17)
distaff　卷线杆(1)
distribution lines　分配板(8)
distributor　批发商(48)
ditch　沟渠(56)
dobby weaves　多臂组织(指多臂织机可以织造的织物组织，小花型组织)(30)
doff　落卷，落筒(14)
doffer　道夫(24)
domestics　家用织物(30)
dope　纺丝液(8)
double breasted　双排纽扣的(44)
double faced structure　双面织物(34)
double needle overlaps structure　重经(36)
draft plan　穿综图(33)
drawing plan　穿综图(33)
draft　牵伸(14)
drainage　污水；排水(3)
drapery　帷幕，悬挂织物(5)
drawing frame　并条机(14)
drawing　牵伸(7)
DREF　德雷夫纺纱法，尘笼纺(16)
dress shield　防汗衬布(5)
dressy　衣着讲究的(43)
drill　斜纹布，卡其(28)
drop wire　经停片(27)
drum warping　分条整经(26)
drum　集聚罗拉；筒管(15)
dry rubbing resistance　耐干摩擦(47)
dry spinning　干法纺丝(8)
dry-laid　干法成网(39)
duct　管道(27)
duplicated creel　双联整经筒子架(26)
durability　耐久性(49)
durable crease　耐久皱褶(4)
durable-press　耐久压烫(6)
dusting off　落浆(27)
dyeability　染色性能(9)
dyed　染色的(1)
dyestuff　染料(9)
ear loops　耳绳，耳带(40)
earthwork　土木工程(56)
easy-care　免烫(6)
effluent　废液(9)
electrets　驻极体(40)
electrification　带电；充电(50)
electromagnetic shielding　电磁屏蔽(12)
electrospinning　静电纺丝(11)
elongation　伸长(18)
embankment　筑堤(41)
embroidery　刺绣(43)
emulate　相仿，与……竞争(9)
emulsion system　乳化洗毛系统(23)
enact　颁布(48)
energy harvesting　能源采集(12)
enforcement　实施，执行(48)
entity　实体(39)
environmentally friendly　环境友好地(3)
erosion　腐蚀，侵蚀(56)
escherichia coli　大肠杆菌(11)
esthetic appeal　美的外观(38)
even twill　双面斜纹(32)
exempt　免除(48)
exhalation valves　呼吸阀(40)

expectancy　期望值(53)
expedit　加快，加急(51)
exponentially　指数级地(1)
extension percentage　伸长率(18)
extrude　挤出(8)
fabric count　织物经纬密度(32)
fabric take-away mechanism　织物牵拉机构(35)
fabric　织物，布，面料(1)
face loop stitch　正面线圈(35)
face masks　口罩(40)
Fahrenheit　华氏温度(51)
faille　罗缎，菲尔绸(32)
failure　破坏(56)
family　(植物的)科(7)
fancy　花式的，风轮(24)
fashion style　时装款式，样式，风格(43)
feeder　喷丝头(49)
fell　织口(29)
felt　毡制品；把……制成毡(39)
felting　毡化(4)
feminine　女性的，妇女的(41)
fiber displacement　纤维位移(46)
fiber migration　纤维迁移(15)
fiberglass　玻璃纤维，玻璃丝(1)
fiber　纤维(1)
fibrillated film　膜裂(39)
fibroin　丝心蛋白(5)
fibrous　纤维状的(7)
fillet　钢丝针布(24)
filling carrier　载纬器(31)
filling-faced twill　纬面斜纹(32)
filling materials　填充物(48)
fill　填充土(56)
filter　过滤，过滤器(1)
filtration　过滤(51)
finish　织物的整理(4)
finisher-drawing　并条机(14)
first responde　现场急救人员(40)
flame retardancy　阻燃(52)
flame retardant textile　阻燃纺织品(38)
flame retardant　阻燃，阻燃剂(38)
flame-retardant　阻燃(10)
flammability　阻燃性(40)
flammable fiber　可燃纤维，易燃纤维(13)
flanged　带凸缘的，用法兰连接的，折边的(29)
flannel　法兰绒(32)
flared skirt　喇叭裙(43)
flashspun　闪蒸(39)
flax　亚麻(1)
flex　屈伸(10)
flexibility　挠性(2)
flexible rapier　挠性剑杆(31)
float　浮长线(32)
floral pattern　花卉图案(30)
flowchart　流程表　(8)
fluffy　毛茸茸的(13)
fluted roller　沟槽罗拉(6)
foam bonding　泡沫黏合(39)
forced air　高压气流(14)
forceps　镊子(13)
foreign fiber　异纤(2)
formaldehyde content　甲醛含量(47)
formula　配方，处方(49)
freight　货物，货运(55)
friction drum　摩擦辊(25)
friction spinning　摩擦纺纱(16)
front rest　胸梁(29)
front shed angle　前部梭口角(30)
FTC(Federal Trade Committee)　美国联邦贸易委员会(48)
full-handling　丰满手感(22)
functional finish　功能整理(38)
functional part　功能零件(55)
fungus　真菌(2)

fuse　熔接(39)
fuzzy　毛茸茸的(17)
gabardine　华达呢(28)
garment　服装(2)
gas dryer　煤气烘干机(2)
gauge　机号(34)
gelation　凝胶化(50)
general finish　一般整理(38)
generic　(生物的)属(48)
Georgia　格鲁吉亚(1)
gill　针梳(22)
gingham　方格色织布(32)
giver　递纬剑，送纬剑(31)
glacial acetic acid　冰醋酸，冰乙酸(9)
glass fiber　玻璃纤维(49)
glazing　轧光(42)
glycol　乙二醇(10)
goggles　护目镜(40)
golf club shaft　高尔夫球杆(49)
gowns　手术衣，隔离衣(40)
graphite fiber　石墨碳纤维(49)
grayish color　泛灰的颜色(7)
greige　坯布，坯绸(38)
grey goods　坯布(38)
gripper guide　导梭片(31)
gripper loom　剑杆织机，片梭织机(31)
grippers　夹子，梭夹(31)
guide roll　导纱辊(27)
gummy　黏稠的(7)
hackle　梳麻(6)
hairiness　毛羽(17)
Han Chinese clothing　汉服(1)
hand　手感(5)
handkerchiefs　手帕(33)
handloom　手工织布机(1)
Hanfu　汉服(1)
hank　纱绞(18)
harness frame　综框(29)
harsh hand　粗糙的手感(38)
head　圈头，针编弧(34)
headstock　车头(26)
heald　综框(29)
healthcare workers　医护人员(40)
heated cylinder　热风烘筒(27)
heat-resistant　耐热的(1)
heavy "ducks"　重型机械(33)
heddle　综丝(27)
helical　螺旋状的(50)
helmets　头盔(57)
hemp　大麻(7)
Hemudu culture　河姆渡文化(1)
herringbone　人字斜纹(32)
hessian　打包麻布(7)
high-wet-modulus　高湿模量(9)
hinder　阻碍，妨碍，阻挡(19)
home furnishings　家庭装饰(2)
homespun　钢花呢，手工纺织呢(32)
homogeneous　均匀的，同质的(24)
hook　针钩(36)
hook　提花机竖钩(30)
hopper　料斗，棉箱(14)
Hopsack　方平组织(33)
horny　角状的(4)
horticulture　园艺学(52)
hose　软管，水龙带(49)
Huck　浮松(33)
hue　色彩(3)
human civilization　人类文明(1)
hydrocellulose　水解纤维素(23)
hydroentanglement　水刺(39)
hydro-extractor　离心脱水机(23)
hydrogen bond　氢键(2)
hydrogen peroxide　过氧化氢(5)
hydrogeology　水文地质(52)
hydrophilic fibers　亲水纤维(46)
hydrophilic　亲水性的(2)

hydrophobic　疏水的，拒水的(3)
hydroxyl groups　羟基(2)
hygiene　卫生学(41)
hygroscopic　吸湿的(4)
hygroscopicity　吸湿性(27)
hypochlorite　次氯酸盐(9)
identification　鉴别(13)
ignite　点燃，使燃烧(38)
immense　巨大的，广大的(50)
immersion roll　浸没辊(27)
immiscible　互不相溶的(50)
impart　给予(2)
impervious　不可渗透的(41)
incandescent bulbs　白炽灯(8)
incineration　焚化，垃圾焚烧(57)
incontinence　失禁，无节制(41)
index of birefringence　双折射率(13)
indirect take-up　间接卷取(29)
industrial nylon　工业用尼龙(49)
Industrial Revolution　工业革命(1)
industrial textiles　产业用纺织品(49)
infant products　婴幼儿产品(47)
inflammation　发炎(11)
influenza　流行性感冒(40)
infrared spectra　红外光谱(9)
infrared spectrophotometer　红外分光光度计(13)
infrared　红外线(27)
infrastructure　基础设施(53)
initial modulus　初始模量(49)
instantaneous　瞬间的，即刻的，即时的(49)
intact　未经触动的(6)
integral　完整的，整体的；构成整体所必需的(50)
interfacial polymerization　界面聚合(50)
interfacial　界面的(4)
interfacing　界面(42)
interior trim　内饰(52)
interior　内部，内部装饰(55)
interlace　交织(29)
interlining　中间衬料(48)
interlock　双罗纹(36)
intermediate　二道梳毛机(24)
intermesh　相互穿套(34)
intermittent system　间歇式纺纱系统(14)
International Wool Secretariat(IWS)　国际羊毛咨询处(38)
interoperability　互通性，互操作性(47)
interstice　空隙(2)
iron　熨烫(2)
isotropy　各向同性(39)
jacket　夹克(17)
Jacquard head　提花龙头(30)
Jacquard loom　提花织机(1)
jean　粗斜纹棉布，三页细斜纹布(32)
jersey　针织套头衫(10)
jute　黄麻(7)
kemp　死毛，戗毛(4)
keratin　角蛋白(5)
Kevlar　凯夫拉芳纶(49)
kink　扭结(18)
knit　针织(1)
knitting　针织(34)
knitting cycle　编织循环(35)
knitting system　编织系统(35)
knocking-over　脱圈(35)
labor-intensive　劳动力密集型的(7)
landfill　垃圾填埋(57)
Lanital　(用酪素纤维制成的)人造羊毛(9)
lapper　成网机，成网机构(14)
latch　针舌(35)
latch needle　舌针(35)
lateral force　侧向力(31)
lattice　输送帘子(14)

laundry aids　洗涤剂(6)
lay　筘座(29)
leakage　泄露(41)
left-hand twill　左斜纹(32)
legislation　法规(48)
legs or side limbs　圈柱(34)
leotards　紧身连衣裤(10)
let off　送经(27)
letting-off　送经(29)
Liangzhu culture site　良渚文化遗址(1)
lifting plan　提综图，纹板图，也叫 peg plan(33)
lightweight　轻质的(1)
lignin　木质素(3)
limp　松软的(9)
line　长纤维，长麻(6)
linear velocity　线速度(25)
linen　亚麻(6)
lingerie　女内衣(10)
lining fabric　衬里织物(32)
lining　衣服衬里(17)
linoleum　漆布(7)
lint　棉绒(28)
literal　实际的(56)
litmus　石蕊(试剂)(47)
little more than　差不多，仅仅是(8)
lobated　叶状的，分裂状(8)
lock　毛撮(24)
loft　蓬松的，高雅的(4)
longevity　寿命(27)
loom beam　织轴(27)
loom　织机(1)
loop　线圈(34)
lotus leaf　荷叶(50)
louse　虱子(1)
lubricant　润滑剂(27)
lumen　腔(2)
lunar landing module　月球着陆模块(51)
Lyocell/Tencel　莱赛尔/天丝(9)
macroporosity　大孔隙度；宏观孔隙度(50)
magazine creel　复式筒子架(26)
mainstay　主要力量(2)
making-up　裁剪，缝制(39)
mandate　授权，命令(41)
manganese　锰(12)
manipulate　操纵，操作；巧妙地处理(50)
market niches　细分市场，nich，微环境(9)
matching　拼毛，将同级套毛并在一起(23)
matt　不光亮的；无光泽的；亚光的(19)
matt　方平组织(33)
mechanism　机理(9)
medieval　中世纪的(18)
medulla　毛髓(4)
meltblown　熔喷(39)
melting point　熔点(13)
mercerize　丝光处理(3)
methodology　方法学，方法论(13)
microorganism　微生物(7)
midi　半长裙(44)
mildew　发霉(2)
milled cloth　缩绒织物，缩呢织物(22)
millennia　千年(1)
mineralizer　矿化剂；造矿元素(50)
mini style　服装的超短款式(44)
minimal handling　尽量缓和的操作(5)
miniskirt　超短裙(43)
miscellaneous　杂乱的(52)
mode　模式，方式(9)
moderately　中等的(2)
modulus　模量(49)
moisture absorption　吸湿性(46)
moisture regain　回潮，回潮率(4)
moisture vapour transport　透湿性(19)
mold　霉菌(2)

molecular orientation　分子取向(9)
moleskin　仿鼹鼠皮，鼹鼠皮(32)
monofilament　单丝(46)
mood　服装的情调(43)
morphologically　形态方面(9)
motif　花纹图案(30)
mule　走锭纺纱机(22)
multiaxial　多轴向(52)
multiple package creel　复式筒子架(26)
multiplicative tensioner　倍积式张力器(25)
mungo　硬再生毛(22)
muslin　平纹细布，薄纱织物(32)
nanocrystal　纳米晶体(50)
nanodroplet　纳米液滴(50)
nanofillers　纳米填加物(57)
napkin　餐巾(30)
native cellulose　天然纤维素(9)
necessitate　需要(7)
necklace-like　项链状(50)
needle loop　圈干(34)
needle　针(1)
needlepunching　针刺(39)
negative let-off　消极式送经(29)
neps　棉结(19)
nettle　荨麻(7)
neutralize　中和(7)
nitrocellulose　硝化纤维，硝化纤维素(8)
noil　落毛，落纤维(22)
no-iron　免烫(2)
nomenclature　名称，术语(18)
non-stationary single end creel　移动单式筒子架(26)
non-fibrous　非纤维状的(7)
nonwovens　非织，非织造布(1)
norm　标准(28)
nose strips　鼻夹(40)
nozzle　喷嘴(15)
nucleation　成核现象；集结(50)
nutrient　营养物，滋养物(7)
oceanographic cable　深海缆绳(49)
odor　气味(13)
off round　偏离圆形(9)
olefin　烯烃(7)
olefins　聚烯烃(8)
open lap　开口线圈(34)
open-end spinning　自由端纺纱，气流纺纱(15)
opener　开棉机，开松机(14)
opening　开松(14)
organdy　蝉翼纱(32)
organic fertilizer　有机肥(3)
organic solvent　有机溶剂(2)
orifice　外孔，小孔(49)
originated　起源(1)
ottoman　粗横棱纹织物(32)
over end withdrawal　轴向退绕(25)
overlap　针前垫纱，圈干(34)
overtwist　加捻过度(17)
package build　卷装成形(25)
package holder　筒子座，卷装握持器(26)
pad　衬垫(49)
pajamas　睡衣(17)
pants　裤子(43)
pantyhose　连裤袜(10)
para-and metha-types　对位与间位(49)
para-aramids　对位芳纶(57)
parachutes　降落伞(51)
passenger coach　长途公共汽车(55)
passenger vessel　客船(55)
passer　检查员(23)
pathogen　病原体，病菌(40)
pattern chain　纹链(30)
pattern drum　分条整经大滚筒(26)
pattern row　花型横列(34)
pattern　花型(29)
patterned fabric　花纹织物(30)

pectin 果胶(3)
pedal pushers (长及小腿的)女运动裤，自行车裤(43)
peer 同等的人(44)
pendant 悬垂的(50)
perborate 过硼酸盐(5)
percale 高级密织薄纱(32)
perception 知觉，感知，洞察力(8)
perennial 多年生的(7)
perforated apron 带孔的胶圈(15)
perforated drum 集聚罗拉(15)
permeable 透水的(52)
permittivity 介电常数(19)
peroxide bleach 过氧化物漂白剂(9)
pertain 关于(48)
pesticide 杀虫剂(3)
phase separation 相分离(50)
phloem 韧皮(6)
photoanode 光阳极(12)
photomicrograph 显微照片(13)
physico-chemical 物理化学的(11)
physiological 生理的(46)
pick 纬纱(29)
picker lap 清棉棉卷(14)
picker 清棉机(14)
picking unit 清棉装置(14)
picking 投梭，引纬(29)
pick-spacing 纬纱间距(29)
piezoelectric 压电(12)
pile 堆，绒头(10)
pillar stitch 编链(36)
pillowcase 枕套(2)
pin 棒针(34)
piqué 凹凸组织(30)
pitch 场地，这里指沥青(49)
plain 纬平针(36)
plain shuttle loom 平纹织物织机(29)
plain weave 平纹织物(29)
plant fibers 植物纤维(1)
plantation 农场(3)
pleat 皱褶，打褶(4)
pleated skirt 百褶裙(43)
pliable 柔顺的(2)
plugging 堵塞(52)
ply 合股(49)
polarizing microscope 偏振光显微镜(13)
poly(tetrafluoroethylene) 聚四氟乙烯(12)
polyacetylene 聚乙炔(12)
polyamide 聚酰胺(57)
polyaniline 聚苯胺(12)
polycarbonate 聚碳酸酯(40)
polydispersity 多分散性(50)
polyester/cotton blend 涤/棉混纺(38)
polyester 涤纶(2)
polyethylene 聚乙烯(40)
polymer 聚合物(4)
polymer-laid 聚合物直接成网(39)
polynosic fiber 丽赛纤维，虎木棉，波里诺西克纤维，属再生纤维素纤维的一种(9)
polypropylene 聚丙烯(41)
polypyrrole(PPy) 聚吡咯(12)
polystyrene 聚苯乙烯(40)
polyurethane 聚氨酯(41)
polyvinyl alcohol(PVA) 聚乙烯醇(27)
pond underliner 池底铺层(41)
pool retting 池塘沤麻(6)
poplin 府绸(32)
porogen 致孔剂(50)
porosity 多孔性(38)
positive let-off 积极式送经(29)
postulated 被假设的，假定(50)
post 立柱(25)
precipitation 沉淀(50)
precursor 母体，前驱体(49)
predate 早于(1)

preform　预制件(52)
prehistoric　史前的(1)
primary wall　初生胞壁(2)
print bonding　印花黏合(39)
profile　简介，特质(9)
progression　飞数(32)
progression　进展，发展(57)
projectile guides　导梭片(31)
projectile loom　片梭织机(31)
prominent　突出的，显著的(2)
propulsive zone　推进区，牵引区(31)
protective clothing　防护服(40)
prototype　原型，样板(52)
protrude out　突出(19)
protruding　突出的(15)
proximity　接近(33)
pulling stress　拉伸应力(45)
pulp　浆粕(9)
pulverize　粉碎(23)
punched cards　纹板(30)
purl　双反面(36)
purport　似乎(48)
qualitative　定性的(13)
queen of fibers　纤维皇后(5)
quilting　被子(33)
radioactive　放射性的(41)
radiofrequency　射频(11)
rag　碎布(10)
rag-tops　抹布(51)
ramie　苎麻(7)
rapier head　剑头(31)
rapier loom　剑杆织机(31)
rate of weft insertion　入纬率(31)
rayon　人造丝(9)
reactant　反应物(9)
reclining twill　缓斜纹(32)
recreational　休闲的(2)
recycled wool　再生毛(48)
recycling　循环利用(57)
redeposition　再沉积(10)
reed　钢筘(27)
reel　卷(9)
reeling　缫丝，络丝(5)
refractive index　折射率(13)
refractive indices　折射率(13)
regenerated cellulose　再生纤维素(9)
regular figure　规则花纹(30)
regular twill　正则斜纹(32)
reinforce　加强，加固(1)
relaxation shrinkage　松弛回缩(39)
reservoir　储液器，储液装置(50)
residue　残渣(13)
resin finish　树脂整理(6)
resin　树脂(49)
respirators　呼吸器，呼吸机，呼吸面罩(40)
rest position　起始位置(35)
retting　沤麻(6)
reverse loop stitch　反面线圈(34)
rewind　重新卷绕(25)
rib　罗纹(36)
rib fabric　凸条纹织物(32)
rice-grain　稻粒(50)
ridge　凸棱(17)
right side　织物正面(32)
right-hand twill　右斜纹(32)
rigor　苛刻(27)
ring spinning　环锭纺纱(14)
ring　钢领(14)
ring-spun yarn　环锭纱(14)
rinsing　清洗，漂清(5)
rip　剥(6)
roller　罗拉(14)
rot　腐烂(6)
rotor spinning　气流纺纱(15)
rotor　纺纱杯(15)

roving frame　粗纱机(14)
rug　厚毯(10)
run　针织物脱散(42)
rustling　沙沙响(5)
sacking　麻袋(7)
saliva resistance　耐唾液(47)
sanitary　卫生的(39)
sapphire whisker　蓝宝石晶须(49)
sateen　纬面缎纹(32)
satin　经面缎纹(17)
satin-back crepe　缎背绉(32)
saturation bonding　饱和浸渍黏合(39)
saturation regain　饱和回潮率(6)
saturation　饱和(50)
scaffold　支架(11)
scale　鳞片(4)
scatter　消散(17)
scouring bowl　洗毛槽(23)
scribbler　预梳机(24)
scroop　丝鸣(5)
scutch　打麻(6)
sea vessel　船(55)
sealing　黏结(45)
secondary wall　次生胞壁(2)
section beam　分批整经轴，分条整经轴(26)
section　经纱条带(26)
self-assembly　自组装(50)
Selfil yarn　加长丝自捻纱(16)
self-twisted yarn　自捻纱(16)
selvage　布边(29)
selvedge　边缘(15)
semi-positive let-off　半积极式送经(29)
semi-worsted　半精梳(22)
sericulture　养蚕业，蚕丝业(5)
serrated　锯齿状的(9)
serviceability　耐用性能(6)
servo motor　伺服电动机(29)
set mark　开车痕(28)
setting tank　沉淀桶，澄清桶(23)
sew　缝纫，缝(1)
shantung　山东绸(32)
sharkskin　鲨皮布，鲨鱼皮革(32)
sheaves　(sheaf 的复数形式)捆，束(6)
shed crossing point　综平点(29)
shed soil　阻挡尘土(17)
shed　梭口(29)
shedding　开口(29)
shedding cam　开口踏盘，开口凸轮(30)
shedding cycle　开口循环(29)
shedding　开口(27)
sheen　光彩，光泽(6)
sheet　床单，被单(1)
shirt waist　女用衬衫(43)
shirting madras　衬衫布(30)
shoddy　软再生毛(22)
shoulder pad　垫肩(44)
shredder　破碎机，碎纸机(9)
shrink　收缩(42)
shrink-resistant finish　防缩整理(2)
shrub　灌木(7)
shuttle　梭子(1)
sieve　筛网(15)
silhouette　侧面影像，服装轮廓(43)
silk chiffon　薄绸(43)
Silk Road　丝绸之路(1)
silkworm　蚕(5)
silky　像蚕丝一样的(4)
silk　蚕丝(1)
siltation　淤泥(41)
single breasted　单排纽扣的(44)
single faced structure　单面织物(34)
single package creel　单式筒子架(26)
single　单纱(14)
sinker loop　沉降弧(34)
size box　浆槽(27)

size liquor　浆液(27)
size recipe　浆料配方(27)
sizing　上浆，浆纱(49)
ski　滑雪板(49)
skiwear　滑雪服装(41)
slasher　浆纱机(27)
slashing　上浆(26)
slick　滑溜溜的(17)
slipper satin　鞋面花缎(32)
sliver　条子(6)
slub catcher　清纱器(25)
slubbing　头道粗纱(22)
slub　纱线粗节(25)
snag　钩丝(17)
snick blade　清纱板(25)
sodium carbonate　碳酸钠(6)
sodium hydroxide　氢氧化钠(6)
sodium niobite　铌酸钠(12)
softener　柔软剂(9)
solar cell　太阳能电池(12)
solubility　溶解法，溶解(13)
solvent　溶剂(27)
soundness　羊毛的强力程度(23)
space shuttle　航天飞机(49)
space suits　宇航服，太空服(1)
space vehicle　太空车，宇宙飞船(49)
spandex　斯潘德克斯弹性纤维，氨纶(16)
specialist yarn　特种纱线(55)
specific stress　比应力(18)
spider-web-like　蛛网状(50)
spike　置凸钉(14)
spin　纺纱(1)
spindle　锭子(14)
spindle　纺锤，锭子(1)
spinneret　喷丝头(8)
spinning frame　细纱机(14)
spinning jenny　珍妮纺纱机(1)
spinning machine　纺纱机(1)
spinning mule　走锭纺纱机(1)
spinodal　旋节线；拐点(50)
spontaneously　自发地，自然地；不由自主地(50)
spool　筒子(25)
spray bonding　喷洒黏合(39)
spun yarn　短纤维纱线(2)
spunbond process　纺粘工艺(40)
spunbond　纺粘(39)
spunlace　水刺(39)
spun-laid　纺丝成网(39)
spun-silk　绢丝(18)
squeeze roll　压浆辊(27)
stagnant　不流动的，污浊的(6)
staining test　染色测试(13)
staining　沾色(47)
stalk　茎(6)
staphylococcus aureus　金黄色葡萄球菌(11)
staple　短纤维(2)
starch　淀粉(9)
stationary　固定的(50)
steep twill　急斜纹(32)
steep　泡，浸(9)
stem　针杆(35)
stem　茎(6)
steric　空间的，立体的(50)
stiffness　刚度，硬度(45)
stitch　线圈(34)
stitch density　线圈密度(34)
stitchbond　缝编(39)
stitchbonding　缝编(39)
stitching　针脚，缝合，缝纫(47)
stove　炉子(10)
strain　应变(18)
stray　v. 迷路，偏离，漂泊，漂泊游荡；adj. 迷路的，离群的，偶遇的(17)
stress　应力(2)

striation　条纹(9)
string　线，细绳(49)
stringent　严格的，严厉的(47)
stripper　剥毛辊，剥棉辊(24)
stripping　剥麻(7)
structural materials　结构材料(49)
stuffer yarn　衬垫纱线，填充纱线(30)
subdivide　细分(8)
subjective　主观的，个人认为的(53)
subsoil　下层土(41)
substitute　替代品(39)
substrate　底层(27)
successive　连续的；接连的；相继的(19)
suit　成套衣服(32)
Sulzer　苏尔寿(瑞士公司)(31)
sundries　杂项(10)
supple　灵活的，柔顺的，柔软的(10)
surah　斜纹软绸(32)
surfactant　表面活性剂(50)
surgeon　外科医生(40)
surgical　外科用的(9)
surveillance　监督，监视(57)
susceptible　敏感的(2)
suspend　悬浮(56)
swampy　沼泽的(41)
swift　大锡林，大滚筒(24)
synthetic detergent　合成洗涤剂(5)
synthetic dye　合成染料(1)
synthetic fiber　合成纤维(1)
table covering　桌布(6)
tactile sense　触感，触觉(42)
tactile　触觉的(9)
take up　卷取(27)
taker　接纬剑(31)
take-up roller　刺毛辊(29)
taking-up　卷取(29)
tangential velocity　切向速度(25)
tank retting　池浸沤麻(6)
taping　贴边(7)
targeted　定向的，被定为攻击目标的(50)
tearing strength　撕裂强度(32)
tearing　撕裂(45)
tease　梳理(14)
technical textile　技术纺织品，工业用纺织品，产业用纺织品(55)
temperature regulation　体温调节(11)
temple　边撑(29)
tenacity　强度(18)
tensile strength　拉伸强度(45)
tensile stress　拉伸应力(45)
tentatively　试验性地(13)
tex　特克斯(18)
Textile Fiber Products Identification Acts (TFPIA)　《美国纺织纤维产品鉴定条例》(48)
textile　纺织，纺织物，纺织品(1)
texture　质地(17)
textured yarns　变形纱(9)
texturization　纱线变形工艺(45)
the short-term unevenness　短片段不匀(19)
the spinning triangle　纺纱三角区(15)
the spiral-shaped opening　螺旋形通道(15)
thermal insulation　隔热(52)
thermal properties　热性能(49)
thermal resistance　热阻(46)
thermoplastic　热塑性(52)
thermoset　热固性(52)
thick places　粗节(19)
thick spot　粗节(25)
thin places　细节(19)
thin spot　弱段(25)
thread count　织物经纬密度(32)
thread　线(1)
thready　能明显看出纱线的(22)

tire cord fabric　轮胎帘布(51)
topographical　地形的，地貌的(2)
topography　地形(50)
torsional rigidities　扭转刚度(19)
tortuous　曲折的(52)
tow　短纤维，短麻，落纤(6)
towel　毛巾，手巾(1)
tows　丝束(39)
traveler　钢丝圈(14)
traveling package creel　复式移行筒子架(26)
traversing mechanism　往复导纱机构(25)
tree-like　树状的(50)
tricot lapping　经平(36)
trimming　服装的装饰与修剪(44)
tropical　热带的(56)
trousers　裤子(43)
truck creel　转向筒子架，车式筒子架(26)
tube furnace　管式炉(50)
tubular skirt　筒裙(43)
tuft　簇绒(39)
tumble drying　转笼烘干(5)
tunnel reed　异型筘(喷气织机防气流扩散的构件)(31)
turbulence　湍流(31)
tweed　粗花呢(17)
twill weave　斜纹(32)
twistless yarn　无捻纱(16)
two-way tricot　双向经编织物(10)
ultrasonic bonding　超声波黏合(39)
underlap　针背垫纱，延展线(34)
uneven twill　单面斜纹(32)
uneven　不均匀的(45)
unidirectional　单向的(50)
uniform　制服(41)
unravel　拆散(13)
unsightly　不雅观的(17)
untwist　解捻，退捻(13)
untwist　退捻(18)
unwinding　退绕(25)
upholstery　家具装饰品(10)
vaporization　汽化(46)
vegetable fiber　植物纤维(1)
veil　面纱(49)
vests　背心(57)
viable　可行的(39)
vinyl chloride or vinylidene chloride　氯乙烯/偏氯乙烯(10)
virgin wool　新羊毛(22)
viscose　黏胶，黏液(8)
visual effect　视觉效果(42)
volatile　挥发性的(41)
vulcanize　使硬化，使硫化(24)
wadding　衬垫(10)
waffle cloth　蜂窝纹布，方格纹布(30)
wale　纵行(34)
warp guide　导纱针(35)
warp knitting　经编(34)
warp sizing　经纱上浆(27)
warp stop motion　经纱断头停车装置(29)
warp　经纱(13)
warp-faced twill　经面斜纹(32)
warping　整经(15)
water frame　水力纺纱机(1)
water jet loom　喷水织机(31)
water repellency　拒水性，疏水性(46)
water resistance　耐水洗(47)
water strider legs　水黾腿(50)
water-borne stains　水基污渍(2)
water-splitting　水解(12)
wear　穿(1)
wearable electronics　可穿戴电子产品(57)
weave　机织(1)
weaver beam　织轴(27)
weaving cycle　织造循环(29)
weed killer　除草剂(3)

weft knitting　纬编(34)
weight disc　张力盘(25)
weighting plate　张力盘(25)
weld　焊接(39)
wet spinning　湿法纺丝(8)
wet-laid　湿法成网(39)
wet-milling　湿法缩绒(39)
whip　鞭子；鞭打(50)
whisker　晶须；胡须；腮须(50)
whiskery　须状的(22)
wicking property　芯吸性能(6)
wicking　芯吸(2)
wig　假发(10)
winder　络筒机(25)
winding head　整经机车头(26)
winding　络纱(25)
wipe　擦拭用品(9)
wire card　钢丝梳棉机(14)
wire flat　盖板(14)
woody　木质的(6)
Wool Products Labeling Act(WPL)　《羊毛产品标签法》(48)
wool　羊毛(1)
worker　工作辊，梳毛辊(24)
wrap　一圈，缠绕(25)
wrinkle recovery　褶皱回复，折痕回复(32)
wrong side　织物反面(32)
X-ray diffraction machine　X射线衍射仪(13)
Xinjiang cotton　新疆棉(2)
Yangshao culture　仰韶文化(1)
yank　猛拉(49)
yardage　使用费用(41)
yarn count　纱线支数，纱线密度(18)
yarn number　纱线支数(18)
yarn size　纱线支数，纱线粗细(18)
yarn supply　供纱(35)
yarn　纱(1)
year-on-year growth　同比增长(8)
zig-zag Z字形(33)

References

[1]KUZNETSOV V A, KUZMINA I P, SYLVESTROVA I M. Hydrothermal growth of AIIBVI semiconductors [J]. Bulletin of Materials Science, 1984, 6(2): 177-192.

[2] JOSEPH M L. Introductory textile science[M]. Holt, Rinehart, and Winston, 1981.

[3]ANITA A S. Evaluating apparel quality[M]. Fairchild Publications, USA, 1988.

[4]CHAWLA K K. Fabrous materials[M]. Cambridge University Press, England, 1988.

[5]CARR H, LATHAM B. The technology of clothing manufacture.1994.

[6]JOSEPH M L. Essentials of textiles[M]. Holt Rinehart and Winston Inc. , USA, 1988.

[7]SVEDOVA J. Industrial textiles [M]. Elsevier Science Publishing Co. Inc. , Czechoslovakia, 1990.

[8]GOODWIN J. A Dyer's manual[M]. Pelham Books, Stephen Greene Press, Singapore, 1990.

[9]HATCH K L. , Textile science[M]. West publishing company, New York, 1993.

[10]ADANUR S. Wellington sears handbook of textiles [M]. Technomic Publishing AG, 1995.

[11]SUN X Q. New size and yarn sizing technology [J]. Journal of Zhengzhou Textile Institute, 1999.

[12]MA P X, CHOI J W. Biodegradable polymer scaffolds with well-defined interconnected spherical pore network [J]. Tissue Engineering Part A, 2001, 7(1): 23-33.

[13]SAVILLE B P. Physical testing of textiles[M]. Woodhead publishing Ltd and CRC Press LLC, 2002.

[14]HARTGERINK J D, BENIASH E, STUPP S I. Peptide-amphiphile nanofibers: a versatile scaffold for the preparation of self-assembling materials[J]. PNAS, 2002, 99(8): 5133-5138.

[15] SIMPSON W S, CRA WSHAW G H. Wool: science and technology [M]. Sawston, Cambridge: Woodhead Publishing Limited, 2002.

[16]SHAO Z, VOLLRATH F. Surprising strength of silkworm silk [J]. Nature, 2002, 418(6899): 741.

[17]LAWRENCE C A. Spun yarn technology[M]. CRC Press, USA, 2003.

[18]LORD P R. Handbook of yarn production: technology, science and economics[J]. Crc Press, 2003(9).

[19]AJMERI J R, AJMERI C J. Latest developments in warping technology, 2003.

[20]LI D, XIA Y N. Electrospinning of nanofibers: reinventing the wheel? [J] Advanced materials, 2004, 16(14): 1151-1170.

[21]KESSICK R, TEPPER G. Microscale polymeric helical structures produced by electrospinning[J]. Applied Physics Letters, 2004, 84(23): 4807-4809.

[22]SCHINDLER W D, HAUSER P J. Chemical finishing of textiles. Woodhead Publishing, 2004.

[23]WANG H B, GAO W D. Modern warping technology and its prospects[J]. China Textile Leader, 2004.

[24]杨锁廷. 纺纱学[M], 中国纺织出版社, 2004.

[25]HUANG G. Compressive Behiviours and failure modes of concrete gylinders reinfovced by glass fabric [J]. Materials & Design, 2006(27): 601-604.

[26]ALUIGI A, ZOCCOLA M, VINEIS C, et al. Study on the structure and properties of wool keratin regenerated from formic acid[J]. International Journal of Biological Macromolecules, 2007, 41(3): 266-273.

[27]BROWN P J, STEVENS K. Nanofibers and nanotechnology in textiles[J]. Technical Textiles, 2007.

[28]SPENCER D J. Knitting technology[M]. 宋广礼，李红霞，杨昆，译. 中国纺织出版社，2007.1.

[29]LI L J. Application approach into new winding technology [J]. Textile Accessories, 2007.

[30]RUSSELL S J. Handbook of nonwovens[M]. Woodhead Publishing, 2007.

[31]TRAVIS J. The naked truth? Lice hint at a recent origin of clothing[J]. Science News, 2010, 164(8): 118.

[32]ZHENG Y, GAO X, JIANG L. Directional adhesion of superhydrophobic butterfly wings[J]. Soft Matter 2007, 3: 178-182.

[33]黄故. 纺织英语[M]. 三版. 北京：中国纺织出版社，2008.

[34]MA M, HILL RM, RUTLEDGE G C. A review of recent results on superhydrophobic materials based on micro-and nanofibers [J]. Journal of Adhesion Science & Technology, 2008, 22(15): 1799-1817.

[35]KITTLER R, KAYSER M, STONEKING M. Molecular evolution of Pediculus humanus and the origin of clothing [J]. Current Biology, 2003, 13(16): 1414-1417.

[36]KOLMAKOV, ANDREI. Some recent trends in the fabrication, functionalisation and characterisation of metal oxide nanowire gas sensors [J]. International Journal of Nanotechnology, 2008, 5(4/5): 450-474.

[37]WANG N, CAI Y, ZHANG R Q. Growth of nanowires [J]. Materials Ence & Engineering R, 2008, 60(1-6): 1-51.

[38]BALTER, M. Clothes make the(hu) man [J]. Science, 2009, 325(5946): 1329.

[39]KVAVADZE E, BAR-YOSEF O, BELFER-COHEN A, et al. 30,000-year-old wild flax fibers[J]. Science, 2009, 325(5946): 1359-1359.

[40]JIN Y, YANG D, KANG D, et al. Fabrication of necklace-like structures via electrospinning[J]. Langmuir the Acs Journal of Surfaces & Colloids, 2010, 26(2): 1186-1190.

[41]NAIR A S, YANG S Y, ZHU PN, et al. Rice grain-shaped TiO_2 mesostructures by electrospinning for dye-sensitized solar cells[J]. Chemical communications, 2010, 46(39): 7421-7423.

[42]WANG X, DING B, YU J, et al. Large-scale fabrication of two-dimensional spider-web-like gelatin nano-nets via electro-netting[J]. Colloids and surfaces B, Biointerfaces, 2011, 86(2): 345-352.

[43] KLAESSIG F, MARRAPESE M, ABE S. Current perspectives in nanotechnology terminology and nomenclature[M]. Springer New York, 2011.

[44]ALAGIRUSAMY R. Yarn manufacture-Ⅱ, Indian institute of technology, Delhi, Department of Textile Technology, 2011.

[45]Early history of textiles & clothing. www. Hollings. mmu. ac. uk. Retrieved on 1 January 2012.

[46]GANDHI K L. Woven textiles[M]. Woodhead Publishing Series in Textiles, 2012.

[47]MENDES P G, MOREIRA M L, TEBCHERANI S M, et al. SnO_2 nanocrystals synthesized by microwave-assisted hydrothermal method: towards a relationship between structural and optical properties [J]. Journal of Nanoparticle Research, 2012, 14(3): 1-13.

[48]宋广礼，杨昆. 针织原理[M]. 中国纺织出版社，2013.

[49]NGUYEN L T H, CHEN S, ELUMALAI N K, et al. Biological, chemical, and electronic applications of nanofibers[J]. Macromolecular Materials and Engineering, 2013, 298(8): 822-867.

[50]SUN B, LONG Y Z, ZHANG H D, et al. Advances in three-dimensional nanofibrous macrostructures via electrospinning[J]. Progress in Polymer Ence, 2014, 39(5): 862-890.

[51]VOMIERO A, FERRONI M, COMINI E, et al. Insight into the formation mechanism of one-dimensional indium oxide wires[J]. Crystal Growth & Design, 2014, 10(1): 140-145.

[52] PAUL R. Functional finishes for textiles: Improving comfort, performance and protection [M]. Woodhead Publishing, 2014.

[53] LIN B Q, ZHAO H L. Energy efficiency and conservation in China's chemical fiber industry [J]. Journal of Cleaner Production, 2015, 103: 345-352.

[54] KILINC F S. A review of isolation gowns in healthcare: fabric and gown properties[J]. Journal of Engineered Fibers & Fabrics, 2015, 10(3).

[55] ZHANG R, MA P X. Porous poly(L-lactic acid)/apatite composites created by biomimetic process [J]. Journal of Biomedical Materials Research, 2015, 45(4): 285-293.

[56] HORROCKS R, ANAND S. Handbook of technical textiles [M]. 1st Edition. Woodhead Publishing, 2016.

[57] Bellis, Mary(February 1, 2016). "The history of clothing-How did specific items of clothing develop?". The About Group. Retrieved August 12, 2016.

[58] LI Z, XU Y, FAN L, et al. Fabrication of polyvinylidene fluoride tree-like nanofiber via one-step electrospinning [J]. Materials and Design, 2016, 92: 95-101.

[59] KONCAR V. Smart textiles and their applications (1st ed.) [M]. Duxford, UK: Woodhead Publishing, 2016.

[60] MORAIS D, GUEDES R, LOPES M. Antimicrobial approaches for textiles: From research to market [J]. Materials, 2016, 9: 498.

[61] CHANG H, LUO J, GULGUNJE P, et al. Structural and functional fibers [J]. Annual Review of Materials Research, 2017, 47: 331-359.

[62] CARR D. Forensic textile science [M]. Woodhead Publishing, 2017.

[63] TANG C, MIAO L Y. "Zhongguo Sichoushi" ("History of Silks in China"). Encyclopedia of China [M]. Encyclopedia of China Publishing House, 2017.

[64] LI L, QIN C X, LIU C C. Comparison of the structure and properties of wool and cashmere fibres under potassium permanganate treatment[J]. Fibres and Textiles in Eastern Europe, 2018, 26(4): 29-33.

[65] DOLEZ P. Advanced characterization and testing of textiles[M]. Woodhead Publishing, 2018.

[66] KYUNGHEE P, WONG A Y. Fashion, Identity, and Power in Modern Asia || Woolen Cloths and the Boom of Fancy Kimono: Worsted Muslin and the Development of "Kawaii" Designs in Japan[J]. Fashion, identity, and power in modern Asian popular culture, 2018: 259-284.

[67] DECAENS J, VERMEERSCH O. Specific testing for smart textiles, Advanced Characterization and Testing of Textiles[M]. 2018: 351-374.

[68] TOWNSEND, TERRY. January 2019, Report to the annual meeting of the discover natural fibre initiative, Frankfurt, Germany.

[69] BARHOUM A, PAL K, RAHIER H, et al. Nanofibers as new-generation materials: From spinning and nano-spinning fabrication techniques to emerging applications[J]. Applied Materials Today, 2019, 17: 1-35.

[70] WU Z Y, MCGOOGAN J M. Characteristics of and Important Lessons From the Coronavirus Disease 2019(COVID-19) Outbreak in China [J], JAMA.2020, 323(13): 1239-1242.

[71] HU J, KUMAR B, LU J. Handbook of fibrous materials[J]. John Wiley & Sons, 2020.

[72] SHI Q, SUN J, HOU C, et al. Advanced functional fiber and smart textile[J]. Advanced Fiber Materials, 2019, 1: 3-31.

[73] STEWART C L, THORNBLADE L W, DIAMOND D J, et al. Personal protective equipment and

COVID-19: A review for surgeons [J]. Annals of Surgery, 2020, 272(2): 132-138.

[74] https://www. thoughtco. com/history-of-textile-production-1991659.

[75] https://www. textileschool. com/182/history-of-textiles-ancient-to-modern-fashion-history/.

[76] http://www. silk-road. com/.

[77] https://www. thoughtco. com/textile-machinery-industrial-revolution-4076291.

[78] https://www. iso. org/standards. html.

[79] https://www. qima. com/testing/textile-fabric/importance-textile-testing.

[80] https://www. ntek. org. cn/en/306/170-672. html.

[81] https://www. intertek. com/textiles/gb-18401/.

[82] https://www. intertek. com/textiles-apparel/textile-testing/.

[83] http://www. sjfzxm. com/global/en/426363. html.

[84] http://textilemerchandising. com/classification-of-textile-fibers/.

[85] https://www. textileschool. com/5499/a-new-perspective-to-improve-warp-sizing/.

[86] http://chinuchathome. info/diy-home-shatnez-lab-part-4-how-to-indentify-fabric-fibers. html.

[87] Mary Padron, Clarifying face mask misconceptions crucial as PPE hits mainstream, https://www. ishn. com/articles/112547-clarifying-face-mask-misconceptions-crucial-as-ppe-hits-mainstream, June 7, 2020.

[88] N95 Respirators, Surgical Masks, and Face Masks, https://www. fda. gov/medical-devices/personal-protective-equipment-infection-control/n95-respirators-surgical-masks-and-face-masks, Feb.26, 2021.

[89] China exports over 224 billion masks to assist global COVID-19 fight, http://s. scio. gov. cn/wz/scio/detail2_2021_01/15/2575544. html, Jan.15, 2021.

[90] https://en. wikipedia. org/wiki/Surgical_mask.

[91] https://health. clevelandclinic. org/a-comprehensive-guide-to-face-masks/.

[92] https://en. wikipedia. org/wiki/Medical_gown.

[93] https://www. vizientinc. com/covid-19/isolation-surgical-gown-selection-guide.

[94] https://www. fda. gov/medical-devices/personal-protective-equipment-infection-control/medical-gowns.